ELECTRONICS IN MUSIC

F.C.JUDD

Electronics in Music

Foruli
Classics

Published by Foruli Classics 2012
ISBN 978-1-905792-32-0

First published by Neville Spearman Ltd 1972
This edition published by Foruli Ltd 2012

A CIP catalogue record for this book is available from the British Library

Cover design by Andy Vella at Velladesign (www.velladesign.com)

The original 1972 typesetting has been retained in this edition.
Set in 11 pt Times 2 pt leaded by Ebenezer Baylis and Son Ltd,
The Trinity Press, Worcester and London.

Printed by Lightning Source

Foruli Classics is an imprint of Foruli Ltd, London
www.foruliclassics.com

PREFACE

During recent years music, musical instruments and music reproduction have become greatly influenced by electronics technology, hence the reason for writing this book in which I have attempted to illustrate how and why. As much basic information as possible has been included for the student in music as well as a number of practical circuits for amateur enthusiasts in both electronics and music. It has not been possible however, to provide complete circuits and constructional details for building electronic organs or music synthesizers for example because of their complexity. In any case these are not items that can be successfully constructed without specific technical knowledge and the availability of electronic test instruments. Reference to other publications that deal specifically with electronic organ building for example will be found in the appendix together with details of publications that, from time to time, carry constructional articles on electrical musical instruments and accessories.

The material concerned with electronic music and multi-track music recording (chapters 4 and 5) will I hope, offer something for those with musical talent and a flair for tape recording for both can be combined with highly satisfying results as quite a number of professional musicians have proved. Talents in this direction need not be wasted either because the annual British Tape Recording Contest (see appendix) has a place of entry for amateur efforts.

F.C.J. South Woodford, 1972.

ACKNOWLEDGEMENTS

In compiling a book of this nature the author must of necessity call upon the assistance of manufacturers for information and photographs etc. The following have very kindly and generously contributed:

F. W. O. Bauch Limited (Studio recording equipment)
Dallas Arbiter Limited (Musical instruments)
Dolby Laboratories Inc. (Noise reduction systems)
Dubreq Studios Limited (Electronic Musical instruments)
Electronic Music Studios (London) Limited (Music synthesizers)
EMI Limited (Recording studio equipment)
Grampian Reproducers Limited (Audio equipment)
Grundig (GB) Limited (Tape recorders and hi-fi systems)
Hammond Organ Co. Limited (Electronic Organs)
KEF Limited (Loudspeaker manufacturers)
The Lowrey Organ Company (Electronic Organs)
Metrosound Limited (Hi-fi equipment)
R. A. Moog Limited (Music synthesizers)
Philips Electrical Limited (Audio equipment and electronic organs)
Revox Limited (Tape recorders)
Henri Selmer Limited (Musical instruments)
Thomas Organ Co. Limited (Electronic organs)
Tonus Inc (ARP Music synthesizers)

The author is also indebted to IPC Magazines Limited for permission to reproduce some of his original circuit designs in chapter 3 which were first published in *Practical Wireless and Practical Electronics.*

Thanks are also due to Wout Steenhuis, well-known radio and recording multi-track music artist, Daphne Oram of Oramics Limited, Peter Zinovieff and David Cockerell of EMS Studios (London) Limited (for technical information and photographs etc.), the BBC Radiophonics Workshop and to Desmond Fereday for his excellent line diagrams.

CONTENTS

CHAPTER ONE

Electronics in Music

During the course of the last decade the integration of electronics with music has developed to the stage where computers, electronic 'black boxes' and magnetic tape recorders can be harnessed together to compose, generate and reproduce music with a complex of voices, musical phrasing and dynamic ranges that could never be equalled by a large orchestra of conventional musical instruments. The electronic organ with its attendant electronic percussion-rhythm units and tone cabinets has become both a domesticated and concert hall instrument and the various forms of electric guitar, together with their powerful amplification systems, can produce a volume of sound greater than that from an entire symphony orchestra in full fortissimo. In the home one can reproduce music from radio or gramophone and tape records with fidelity and acoustic spaciousness almost equal to the original.

It is important, however, to first isolate the different applications of electronics in music, for example, electronic organs do not produce electronic music. Instruments of this nature play in tempered scale, as other keyboard instruments and have a given selection of voices and pitch range. The musical sounds are derived from electronic tone generators combined with electronic 'voicing' circuits and so also are sounds produced by electronic percussion-rhythm units. Both require an amplifier and loudspeaker for reproduction. The electric guitar is primarily an acoustic instrument fitted with an electro-magnetic device that converts the natural acoustical vibrations of the strings to electrical currents, hence electric guitar and not electronic guitar. However, for proper reproduction an amplifier and loudspeaker are necessary.

Electronic music on the other hand rather defies explanation because it is not the music that is electronic. It is music derived from electronic tone generators and the 'voices' are created electronically hence confusion by the uninitiated with music played by electronic organs. With the wider range of electronic control at their disposal, composers and creators of electronic music have an almost infinite range of pitch, timbre and dynamics, e.g. they are not bound to tempered scale or to more or less conventional voicing. Such music is normally recorded on magnetic tape, which permits still further changes in the composition should this be necessary, and is then reproduced via one or more amplifying and loudspeaker channels. Multiple channel reproduction systems are frequently used to create various spatial acoustic effects, such as stereophony for instance.

Electronics in sound reproduction have also become pretty sophisticated. High fidelity stereophonic reproduction from radio and disc or tape records is commonplace and certainly provides the listener with an adequate degree of realism. More recent developments include the almost complete elimination of noise from recorded music and a greatly increased degree of control over the acoustic-spatial aspect of reproduction via loudspeakers. Noise, which consists of unwanted continuous hissing, often very audible during quiet passages in music, can be reduced to a very low degree, often as much as 60dB below the wanted sound level. One has only to compare an old 78 rpm electrically recorded disc with a modern LP record to appreciate just how much the background noise on recordings has been reduced. Electronic noise reducing systems used in conjunction with magnetic tape recording can now reduce recording and reproduction noise to a level that can almost be regarded as non-existent. The spatial effect of stereophony has also been considerably widened by development in multi-channel recording and reproducing systems and one claim in respect of such systems is that the acoustics of the concert hall can be re-created in the home.

We have then, three main fields in which electronics play a major part in music—in electronic musical instruments—in electronic music, which might be better regarded as a new form of music—and in the reproduction of music from radio or records. The three have, however, become integrated each with the other. High quality recording and reproducing systems are essential to electronic music and electronic organs are acquiring many of the special electronic techniques used for the creation of electronic music. Equally, electronic music has adopted at least one item common to electronic organs, which is the keyboard, or more correctly the piano keyboard. Here must be mentioned, the music synthesizer, the latest development of the electronics and computer age, because a music synthesizer is perhaps the only self contained instrument that might be classified as an instant electronic music producer. Other chapters in this book deal extensively with electronic music and music synthesizers but some mention of the reason for the invention of music synthesizers may not be out of place at this point.

The creation of electronic music was hitherto a fairly laborious task of first recording all the electronically generated sounds required for a composition and then cutting these from the tape and reassembling them into definite parts or phrases. Many of the reassembled parts were then further modified by electronic treatment and, according to the composer's desire, perhaps even juxtaposed into another order and so on, until the composition was completed. This method made it extremely difficult to produce musical phrases in tempered scale so keyboard controlled tone generators were employed. Most of the earlier forms of keyboard systems were little more than glorified electronic organs and indeed many composers resorted to the use of electronic organs as a keyed tone source. The development of some of the earlier keyboard systems is dealt with later. It soon became necessary to devise a keyboard controlled source of sounds that would not only play tunes in different voices but which also incorporated every conceivable electronic device for generating and shaping sounds far beyond the range of anything that could be created from simple tone generators, electrical filters, modulators and magnetic tape recording techniques. In short, an instant electronic music producer was required. As the chapter dealing with music synthesizers will reveal, composers now have at their disposal a self contained source capable of providing an almost never ending variety of musical voices, continuously variable pitch and dynamic range and complete keyboard control. Moreover the output from these instruments can be connected to a tape recorder or even to an amplifier system for live performance in concert with or without conventional musical instruments which raises the question—is the music synthesizer a new musical instrument? This is debatable but nevertheless derivations of the music synthesizer are appearing in one form or another as attachments to electronic organs to become a second, third or fourth manual, or as small self-contained instruments designed solely for electronically imitating the sounds of conventional instruments such as the piano. Much of the electronic circuitry used for these 'hybrids' has been taken from music synthesizer techniques. But it hasn't ended there. Music synthesizers are also used in conjunction with ordinary (one can't keep calling them conventional) musical instruments, for instance the sound of a piano or even human voices may be taken via a microphone into a synthesizer which can then be programmed to completely change the timbre. The resultant changed voices are normally then recorded onto magnetic tape.

Analysis of Musical Sounds

One could classify all sounds as either musical or as just noise. Sound produced by machinery for example might well be rhythmic but we normally regard it as noise. If one strikes an empty wine glass the ringing sound that is produced might be said to be musical. Almost all sounds do, however, have definable pitch and this includes many 'noise' sounds. Some have the characteristic of a clearly definable pitch plus a noise without audible pitch. A typical example is the combination of the scraping noise and musical note produced by an unskilled violinist with a poorly made violin. In this case the noise and perhaps also the violinist, is unwanted. On the other hand there are musical instruments that produce noise in addition to musical tones but such noise may be nothing to do with the performer. The flute for instance produces an almost pure tone but air noise is also present. Not sufficient perhaps to be distracting to the listener but without the air noise the flute would sound unnatural. A side drum with snare wires produces a sound that is nearly all noise of no particular pitch but combined with the noise is a sound that has definite pitch and this is the 'strike tone' produced by the resonance of the drum head. If the sound of a known musical instrument is to be imitated, or synthesized, then all the components of the sound, including any noise naturally associated with it, must also be generated.

In general it can be said that all musical sounds and many noise sounds have components consisting of pitch, loudness and timbre or tonal quality. These are the characteristics that enable us to distinguish between two notes of the same pitch produced by two different musical instruments or even between two notes of the same pitch and loudness produced by two instruments of the same type, e.g. two violins. In the latter case difficulty might be experienced by those with so called tone deafness but most people with good hearing would be able to distinguish any difference in tonal quality. A trained ear might be well able to separate the harmonics or overtones from the fundamental pitch.

Pitch

The term pitch is normally used in music when referring to the number of vibrations per second at which a musical note is sounding. In electronics the term would be 'frequency, e.g.' one would say that this generator is producing an electrical wave at a frequency of so many Hertz per second. (Note: that the term Hertz, abbreviated Hz, has replaced the former term cycles. Hitherto frequency was expressed in cycles per second or cps.) Pitch can be determined aurally by listening to and comparing a tone of one pitch with the known pitch of another. Those with a trained ear can often quite accurately determine pitch without the aid of a comparison tone. If the pitch of each of two notes is wide apart, including octaves, even an untrained ear can detect the difference, but where two notes are pitched almost in unison with each other, the average ear may not be able to detect a difference in pitch even though it exists. However, when the difference is very small what is called a 'beat note' is produced, which is not normally audible as a distinct tone but as a wavering of one of the two notes being sounded. When the pitch of each note becomes exactly equal the wavering stops, thus indicating about the minimum pitch difference that the ear is capable of recognizing. A skilled piano tuner for instance can usually distinguish between an equally tempered scale and a true scale which involves recognizing a difference of about 1/50th of a semi-tone or something

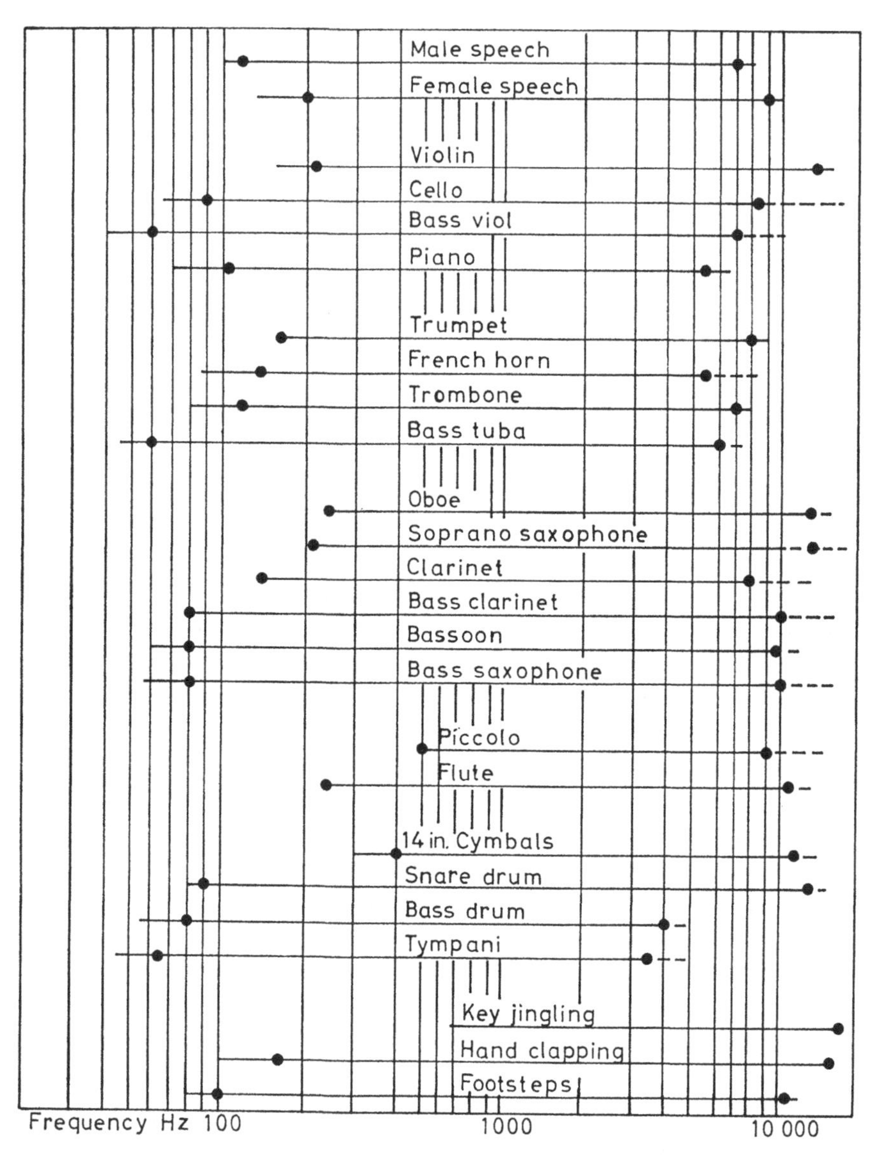

Fig: 1.1

Approximate total frequency ranges of musical instruments, sounds and speech. The ranges shown include fundamentals and harmonics.

like 600 different pitches in an octave. In the region of middle C it is possible to distinguish one note from another when the pitch difference is only 1/120th of a semitone. The ability to do this is normally only due to training an experience. It has been found, however, that the sensitivity of the ear to pitch varies over the normal audio frequency spectrum and that the ear is less sensitive to small changes in pitch at the lower frequencies.

Pitch is sometimes influenced by loudness particularly with pure or sine tones. The pitch of a pure tone may appear to fall as the loudness increases, for example, in the region of 200 to 300Hz an increase in intensity from say 60 to 100 phons may produce an apparent fall in pitch by as much as 10%. In the frequency range 500 to 5000Hz there is little apparent pitch change with increased loudness. Although sensitivity to changes in pitch varies over the normally audible frequency range, the ear can distinguish the pitch of the lowest musical sounds such as those from the double bass. It can also distinguish the highest tones that occur as partials or harmonics and which occur at frequencies as high as 10,000Hz. This is one reason why high quality sound reproducing equipment must have a wide frequency range. The tones in a male voice for instance cover an average range of 120 to 8000Hz, whilst those in a female voice may range from 200 to over 10,000Hz. The chart given in Fig: 1.1 shows the frequency ranges (this includes fundamental pitch range and overtones) of various musical and percussion instruments and speech.

Harmonics of Musical Instruments

Sounds produced by musical instruments are a complexity of the fundamental tone that determines the pitch, a series of harmonics or overtones related to the fundamental, the attack or beginning of the sound, the decay or dying away and any noise which is an inherent production of the instrument concerned. It is the harmonics that determine the characteristic sound or voice of the instrument and it might be said that the attack and decay provide the expression of the voice. The basic sound generators most widely used for musical instruments are strings, air columns, rods, plates and membranes. These are mechanical generators and quite distinct from electronic generators as used in electronic organs. The use of strings as a source of musical sound goes far back into history and as early as the sixth century BC Pythagoras made a study of strings as sound generators. In the seventeenth century a Franciscan Friar, Pere Mersenne, devised the laws that relate to the pitch of a sound produced by a vibrating string to its physical constants. The vibration of strings is a complex subject but Mersenne's laws find application in the function of all stringed instruments. For example pitch is a function of the length of a string and for this reason instruments such as the guitar, banjo and mandolin have 'frets' which are small raised bars along the finger board beneath the strings that enable the player to alter the length of the string and therefore pitch, by a definite amount. With the violin and 'cello, etc., the player has no marked positions for selecting the right notes.

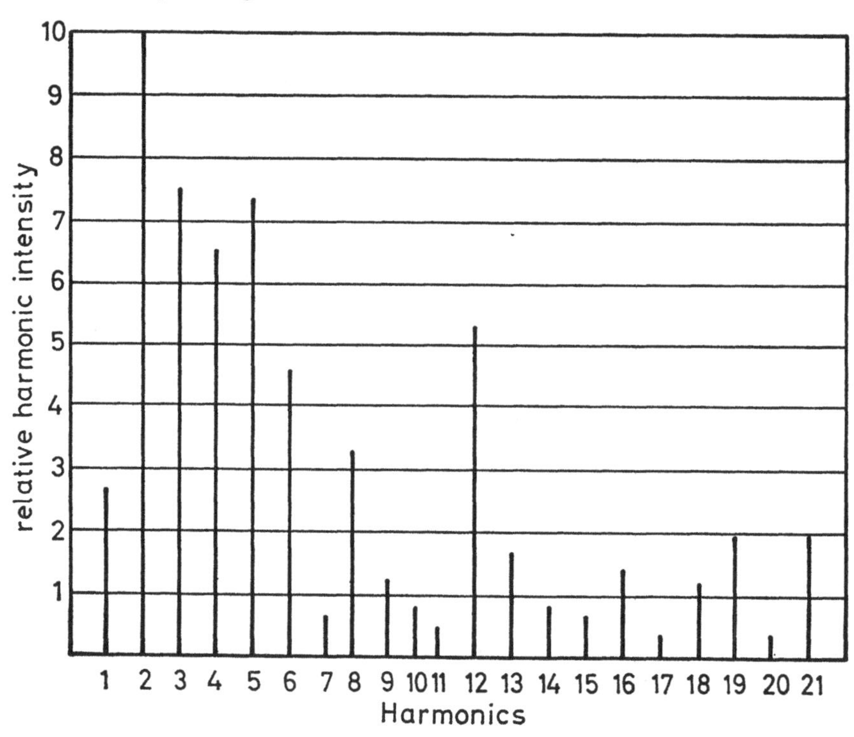

Fig: 1.2
Harmonic spectrums of an open violin G string.

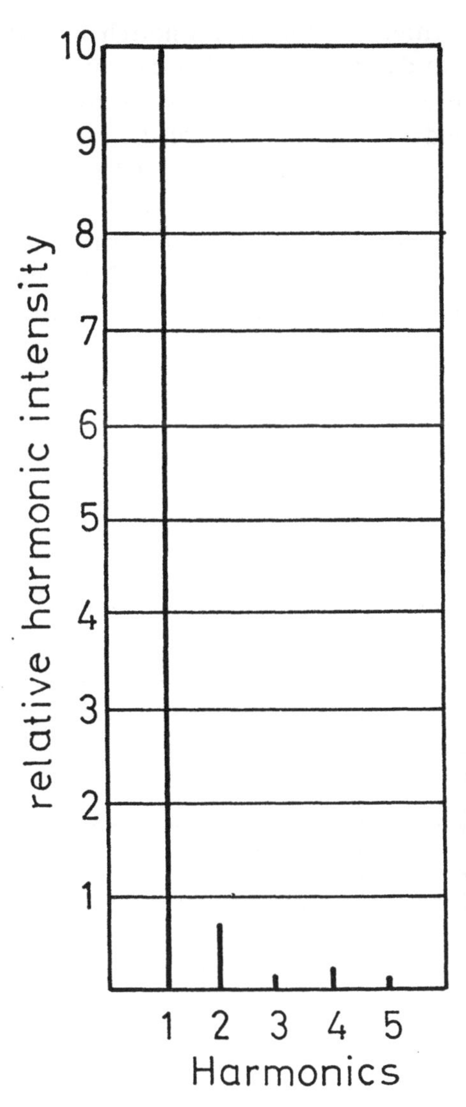

Fig: 1.3
Fundamental and harmonics. Organ flute. Great 4ft A.440Hz.

By suitably exciting a string with a horsehair bow or by striking it, it is possible to develop sounds that contain not only the fundamental but also a series of harmonics, or partials. The order of harmonics depends on the character of the excitation, the point at which this takes place and the density and elasticity of the string itself. The resonating body of conventional stringed instruments also contributes to the nature and intensity of harmonics. The harmonics of a violin open G string, shown in Fig: 1.2, extend up to the 21st but note that the fundamental (1) is not the loudest component.

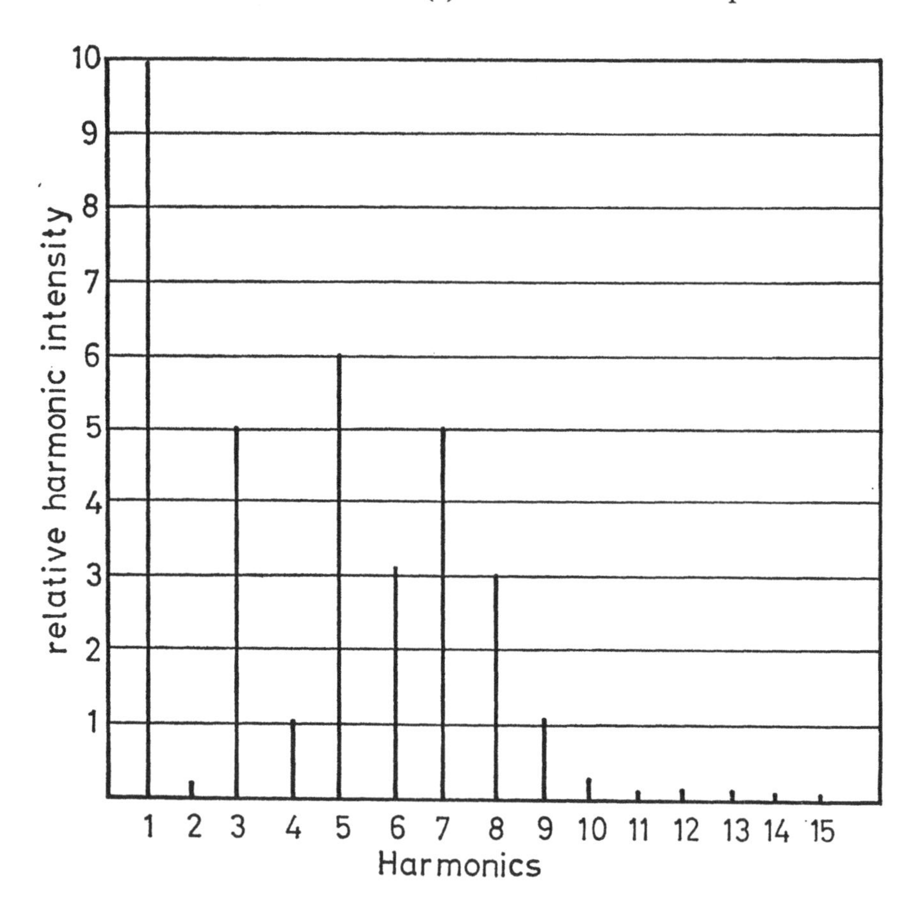

Fig: 1.4
Harmonic spectrum of the B flat clarinet. Fundamental 180Hz.

Wind Instruments

Now let us examine some of the wind instruments of which the organ is probably the most complex since it can be made to imitate the basic voices of other musical instruments. The organ must therefore be capable of producing a wide range of harmonics but on the other hand it can generate a very pure sound almost completely devoid of harmonics as Fig: 1.3 shows. The flute is another wind instrument that produces an almost pure tone but usually has a second harmonic as strong as the fundamental.

Like the flute, most reed instruments are of early origin. The clarinet appears to have evolved from an instrument called the Chalumeau and records indicate that it was first used in 1751 as an orchestral instrument. The clarinet has tonal characteristics similar to all other reed instruments but if the reed is sounded alone (detached from the body of the instrument) it creates a raucous medley of harmonics. Within the body of the instrument the reed is compelled to emit tones corresponding to the natural resonance of the air column. Despite the effect of the resonator however, a large number of harmonics are produced and yet, as shown in Fig: 1.4, some are suppressed, especially the second.

Brass instruments like the French horn, trumpet, trombone and tuba, etc. are not normally classified as reed instruments but do in fact employ a double reed as the primary source of vibration. Brass instruments still produce strong harmonics however, and in the French horn the second and third are predominant and much stronger than the fundamental. The strongest harmonics from the trombone are the second, third and fourth but the second is strongest and equal in intensity to the fundamental.

The human voice has been considered as being almost musically perfect because its variations of pitch, timbre and intensity have yet to be equalled by any musical instrument that man has so far devised. Attempts at synthetic voice production have been made and one notable effort can be heard on the record 'Music from Mathematics' by Brunswick (number STA-8523 stereo or LAT8523 mono). The synthetic voice on this record, which sings to the accompaniment of a synthetic piano, was produced by an IBM 7090 computer coupled to a digital to sound transducer. One of the rather unique features of the human vocal system is that the resonance of the oral and nasal cavities can be consciously controlled and made to behave as a kind of acoustical network. Although the vocal organs are very flexible, any given set of vocal chords and 'acoustical' cavities will have definite tonal characteristics, or in other words a particular 'voice'. The tonal characteristics of a person's voice are as individual as fingerprints. The human voice has been mentioned here simply because it can be synthesized electronically like the sounds of musical instruments, in fact one of the features of modern music synthesizers is that some can create pseudo voices.

Loudness

The loudness of any tone is a function, not only of its intensity but also of its frequency as indicated by the contour curves of equal loudness devised by Fletcher and Munson. The unit of loudness level is the 'phon' and the loudness level of a sound (in phons) is numerically equal to the intensity level in decibels of a 1000Hz pure tone which is judged by the listener to be equally loud. The reference intensity is 10^{-16} watt per square centimetre which is near the value of the threshold of audibility for a 1000Hz pure tone. The phon is little used as a measure of the loudness of sound produced by musical instruments, particularly by electronic or electric instruments whose sound is reproduced by loudspeakers. One is more likely to find the effective loudness of an electronic organ for instance expressed in watts which is the electrical power delivered to the loudspeakers by the final or power amplifier stage. Equally, electric guitarists rate the sound level they can produce by the power output of the guitar amplifier. Relative power and signal voltage levels are more likely to be expressed in decibels (abbreviated dB). Users of music synthesizers for instance may find differences between output signal levels expressed in decibels. The relationship of unwanted noise to signal level is also expressed in decibels. For example a noise level of 60dB below the level of signal voltage means that the actual voltage of the noise is 1000 times below the voltage of the signal level. Although differences in power levels are also expressed in decibels the ratio will not be the same. A noise level given as 60dB below a given signal *power* level means that the power level of the noise is 1,000,000 times less than the power of the signal level.

Another aspect of loudness is the dynamic or intensity level of sounds, musical or otherwise, with respect to each other and here we think in terms of the lowest level of sound that can be detected by human ears known as the *threshold of hearing* to the loudest level that can be withstood before actual pain occurs, the *threshold of pain*, and which occurs when intensity is about 130dB above the threshold of hearing. Within this range are the dynamic ranges of music and speech. The upper level of the music range, probably originally based on the loudness produced by a large orchestra, is about 100dB. Modern high power music amplifiers and speaker systems used by 'pop' music groups produce a loudness level much higher than this and young people constantly subjected to these high sound levels often complain of deafness. This is not surprising in view of the fact that these groups frequently use in excess of 2000 watts of audio power. At close range, in the confines of a small dance hall for instance, the sound intensity may well be close to the threshold of pain.

Vibrato, tremulant and pitch variation

Other elements of musical sound are vibrato which is generally accepted as a rhythmic pitch variation and tremulant, which is a rhythmic loudness variation. Also associated with certain musical instruments is a form of pitch variation generally produced by the method of playing, for instance the slight upward or downward approach to true pitch produced by a violinist or the glissando effect obtained with the Hawaiian guitar. With some musical instruments like the mandolin it is possible to produce a form of tremolo or rapid repetition of the playing of a single note. All these effects can be produced electronically and most are available to the electronic organist, the vibrato and tremulant being the most common.

Sound Generation with electronics

Despite the complexity of musical sounds many, if not all, including the human voice, can be produced by electronic means. With the exception of the electronic organ, musical instrument sounds are

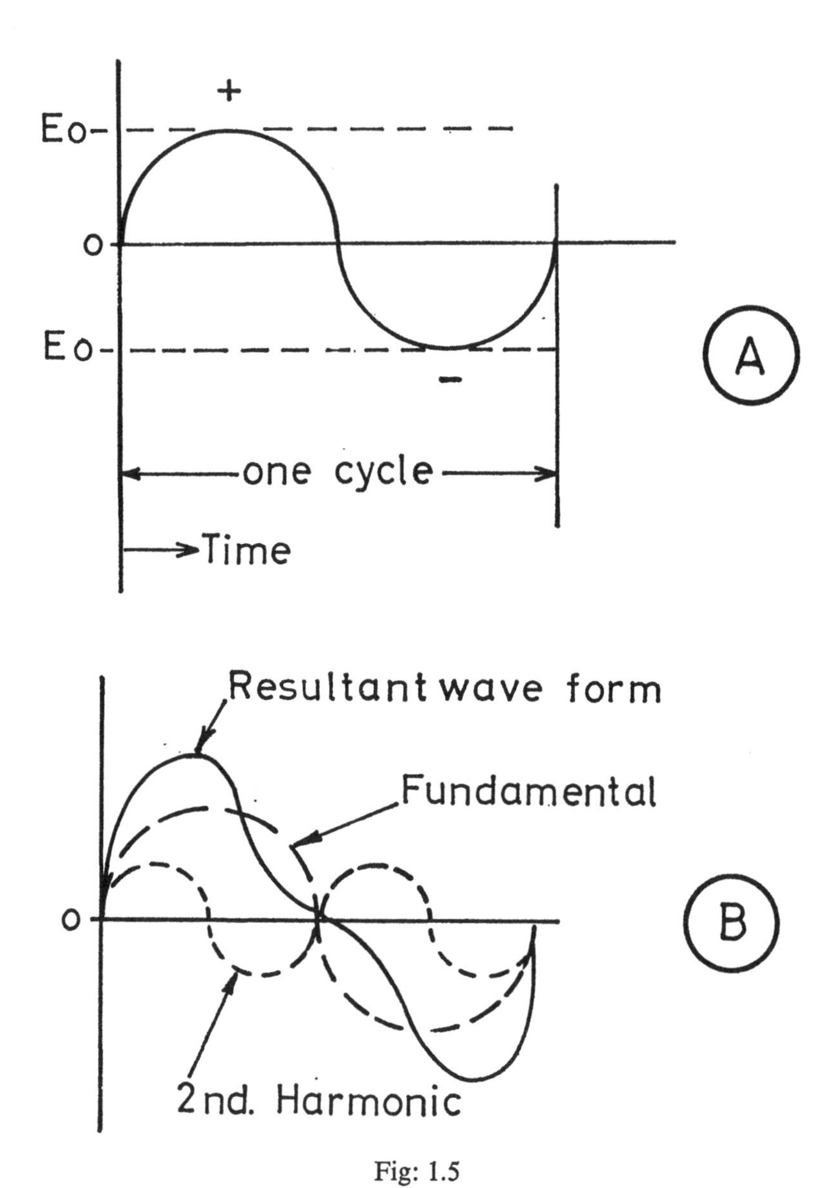

Fig: 1.5
(A) The sine-wave. (B) Effect of adding the second harmonic only.

produced by mechanical means and directly create the air pressure variations that our ears interpret as a sound. Sounds created electronically are first generated as small electrical voltages which are then amplified and applied to a loudspeaker. This responds to the variations in electrical voltages and in turn creates the varying air pressures essential to natural hearing. This process can be reversed, in which case a microphone is used to change the air pressure variations produced by sound into small electrical voltages that can

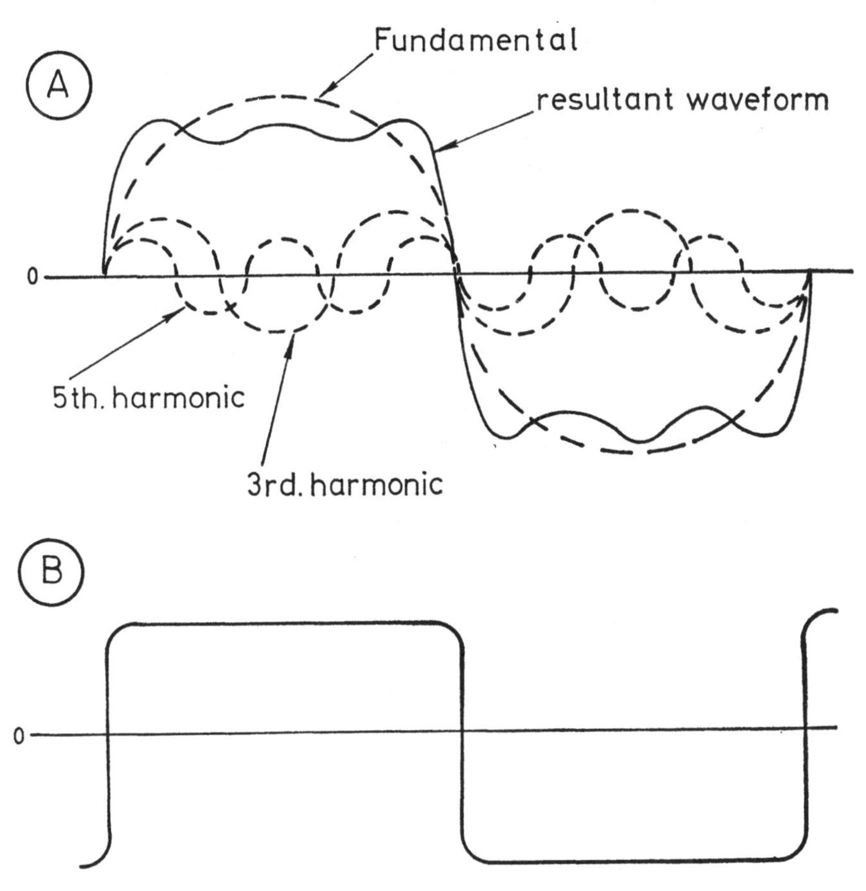

Fig: 1.6

(A) Addition of the 3rd and 5th harmonics to a sine-wave. (B) When enough odd harmonics have been added a square-wave is produced.

be either amplified and reproduced by a loudspeaker at much greater volume than the original or, very conveniently, recorded on magnetic tape for future reproduction. It is not within the scope of this book to deal with the basics of sound reproduction and recording. There are many good textbooks devoted to this already. The rest of this chapter is therefore, devoted to electronically generated sound waveforms and some simple generator circuits.

The fundamental frequency, harmonics, sound envelope (attack and decay) and loudness of a musical instrument all contribute to the production of a varying complexity of air pressures. With electronic generation these characteristics are produced as a complexity of voltage variations which can very conveniently be visually examined by means of an oscilloscope or plotted as graphs or recorded on magnetic tape. The sine-wave is the most common of electrically generated waves—we can call them waves, or waveforms now instead of sounds—and if it is extremely pure, contains no harmonics. It is a re-current wave and its frequency is determined by the number of complete re-currences or alternations per second. A sine-wave voltage alternates between zero and maximum in one direction (positive) and zero and maximum in the opposite direction (negative), as shown in Fig: 1.5A which represents one complete cycle. As the graph shows the voltage rises gradually to its positive peak at E_O and back to zero and then repeats the process in the negative direction. The rate of change with time is sinusoidal and can be expressed by $e = E_O \sin\omega t$ where e is the instantaneous value of the voltage, E_O is the maximum and ω is 2π times the frequency at which the voltage alternates.

If a sine-wave is reproduced via a loudspeaker, the sound, although pure, is not particularly interesting musically. It is only when harmonics are added that a re-current wave of this nature begins to produce a more musically pleasing sound. Continued addition of harmonics will eventually change the form of a sine-wave to one that is either square or triangular (sawtooth). The effect of adding the second harmonic only is shown in Fig: 1.5B. Adding both the third and fifth harmonics as in Fig: 1.6A illustrates how the form of a sine-wave can be changed to take on a square shape, although many more harmonics would be required to produce a perfectly square form as shown in Fig: 1.6B. Such a wave is then called a square-wave and its special property when generated directly by a square-wave generator, is that it contains only odd harmonics in addition to the fundamental. The square-wave is frequently used in the production of electronic music and as a basic waveform in electronic

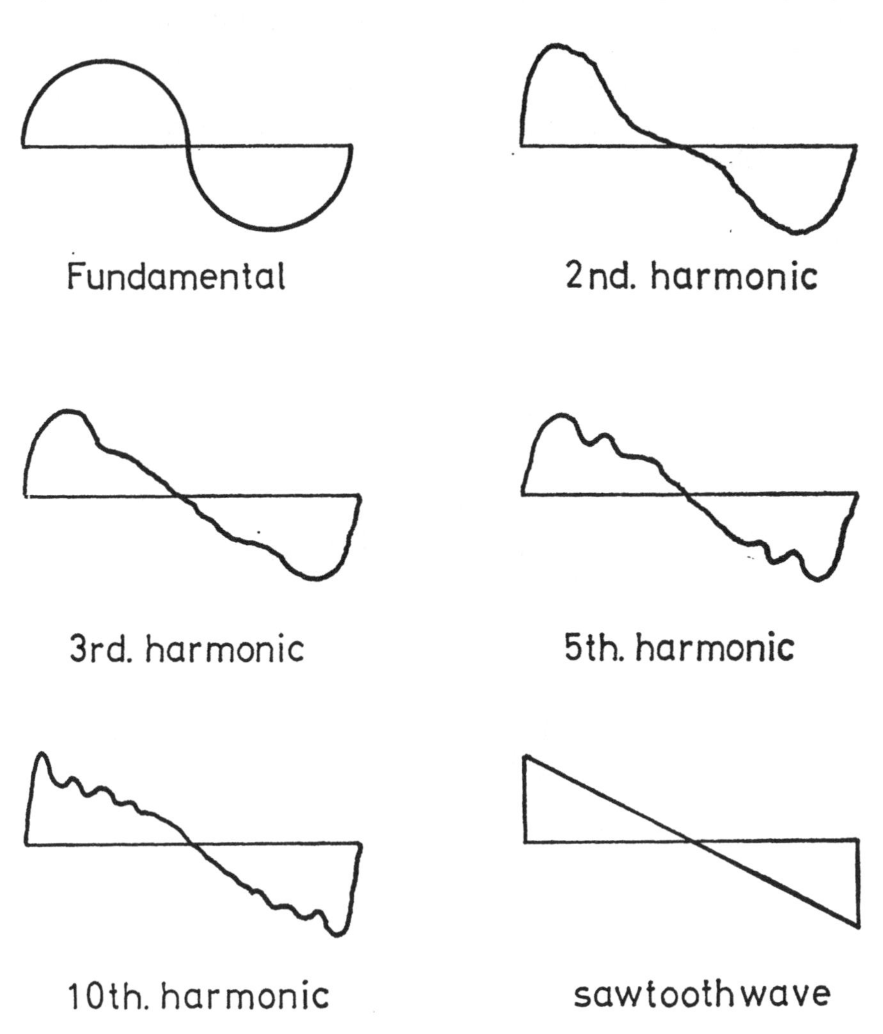

Fig: 1.7
The formation of a sawtooth wave by adding both even and odd harmonics. A good sawtooth wave may contain harmonics up to at least the 30th.

organs chiefly because of its high harmonic content and because it is easy to extract unwanted harmonics from it by electrical filtering.

Another waveform commonly used in electronic music and as a basic tone for electronic organs, is the 'sawtooth' so named because of its shape. If we began with a sine-wave and added both odd and even harmonics the sine-wave would gradually be transformed into a sawtooth as shown in Fig: 1.7. A directly generated sawtooth wave contains harmonics up to at least the 30th and like the square-wave has a rich musical quality.

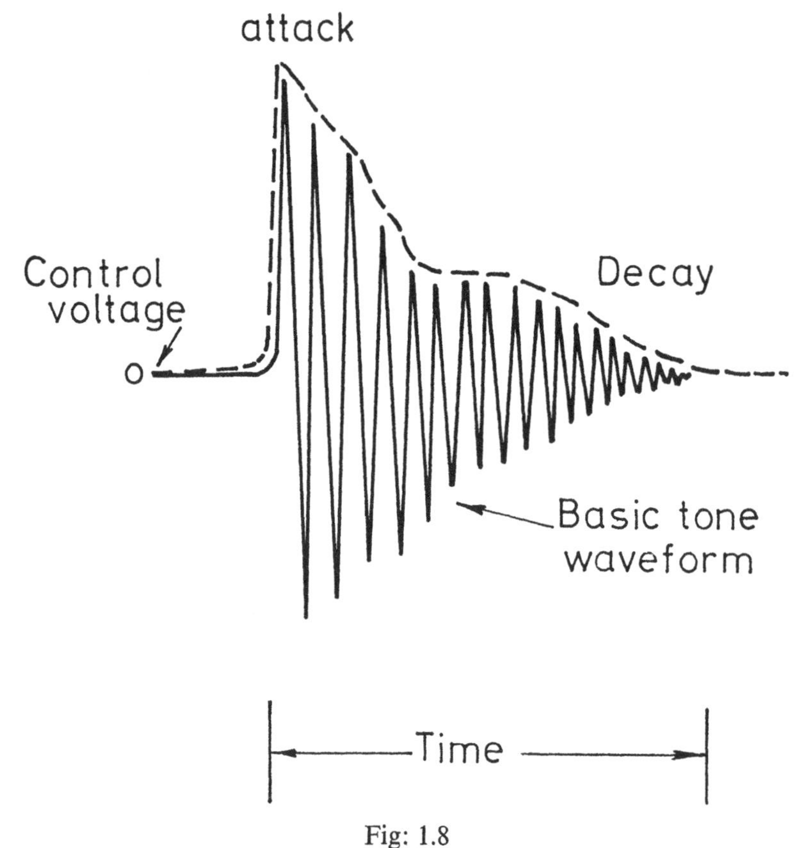

Fig: 1.8
Voltage control is used to produce attack and decay for an electronically generated basic waveform.

With electronics it is possible to generate waveforms of all shapes and at frequencies covering the entire audible spectrum of sound. In this way the basic voices of known musical instruments can be imitated simply by including the correct number of harmonics at their respective intensities. Even noise content associated with the sound can be included because this too can be generated electronically. In order to provide the basic voice waveforms with the correct voice expression, i.e. the attack and decay, further electronic generators are required to provide what is commonly called an envelope for the sound. For example when a piano note is sounded the beginning of the sound is very fast and fairly loud. It then continues for a short period and gradually dies away. What is required now is another generator that will (a) switch on the basic note very quickly and (b) allow the note to sound and then gradually die away. As illustrated in Fig: 1.8 the controlling generator simply produces a waveform that first rises extremely fast to a given level. This would turn on the basic waveform generator and produce the equivalent of the initial striking of the piano string by the hammer, i.e. the attack. The controlling generator wave must now fall to a mean level to maintain the sound (equal to that from the piano string after the hammer has struck) and then gradually fall to zero (dying away of sound). The waveform from the controlling generator may itself be triggered off manually by a key or even by another voltage. The latter method, now commonly used in music synthesizers, is called 'voltage control'.

Electronic tone generators

Prior to the invention of the transistor the generation of recurrent electrical voltages of small amplitude, at frequencies within the audible spectrum, was accomplished with the thermionic radio valve or with thyratrons which are special gas filled valves and which were frequently used as generators of sawtooth waves. Neon bulbs were also used in combination with resistive and capacitive elements to produce waveforms closely resembling the sawtooth. The thermionic valve was perhaps the most flexible in its use as a tone generator because it could be made to produce sine-waves, or square-waves, and with appropriate circuitry, many other and complex waveforms. The main disadvantage of the valve, however, was that it required heater voltage as well as fairly high operating voltages. The valve itself was also relatively large and because of the high operating potentials its associated components and power supplies were much larger than are used now for transistor circuits. As a consequence multi-valve equipment was usually very bulky and heavy. The transistor is not only extremly small but there are a great

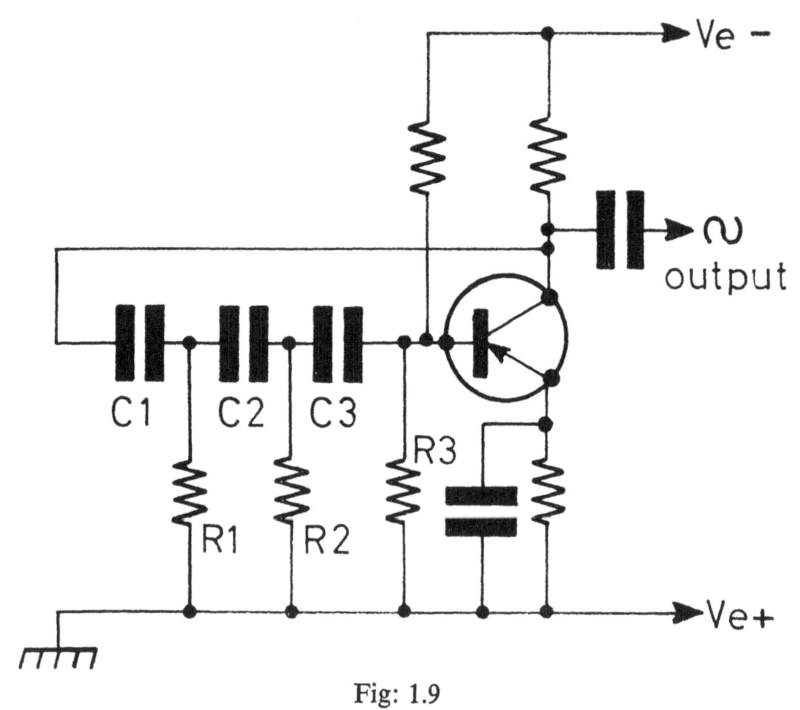

Fig: 1.9
Basic phase-shift oscillator for generating a sine-wave. Frequency determined by C1, C2, C3 and R1, R2, R3.

many different types that enables us to design circuits for functions that would have been difficult and in some cases impossible with valves. Components used in transistor circuits are small and the operating potentials of transistors are quite low. Complex circuits requiring several transistors and their components can be accommodated on circuit boards only a few square inches in size although development in integrated circuitry now permits the inclusion of a whole circuit of transistors and components within capsules often no larger than a single discrete transistor.

The amateur enthusiast who wishes to build transistor circuits does not have to know the theory of semi-conductors, which is what the transistor is, but some knowledge of practical transistor circuitry, experience in arranging layout for transistor circuits and in construction generally is essential if any of the circuits given in this book are to be built and successfully operated. This point is made because it has not been possible to include the vast number of diagrams that would be necessary to show the layout and other constructional details for all the circuits that have been included. Will readers kindly note however, that circuit diagrams used purely to illustrate basic or typical circuits do not contain component values but they may rest assured that those circuits which do specify component values and transistor types, etc., have all been constructed and tested.

Basic oscillator circuits

One of the most simple and perhaps the most commonly used circuits for the generation of sine-waves, is the phase-shift oscillator. It will produce a relatively pure sine-wave at frequencies ranging from sub-audio to supersonic. Normally, generation is at a fixed frequency and this is determined by the phase shift network components shown in Fig: 1.9 as C1, C2, C3 and R1, R2 and R3. A small variation about the nominal frequency can, however, be obtained by a small variation in the value of R1 or R2. The transistors used for phase shift oscillators working within the audio frequency range may be ordinary small signal pnp or npn types. The available peak-to-peak output signal can be as high as 10V depending on the transistor and its operating potential. Such oscillators must, however, not be heavily loaded for this can cause oscillation to stop. It is usual to couple the output to its load through a series resistor to prevent excessive loading. The phase shift oscillator is quite stable and if a stabilized supply voltage is used frequency drift may be as little as one cycle in a thousand.

The inductive oscillator, like the Hartley circuit shown in Fig: 1.10

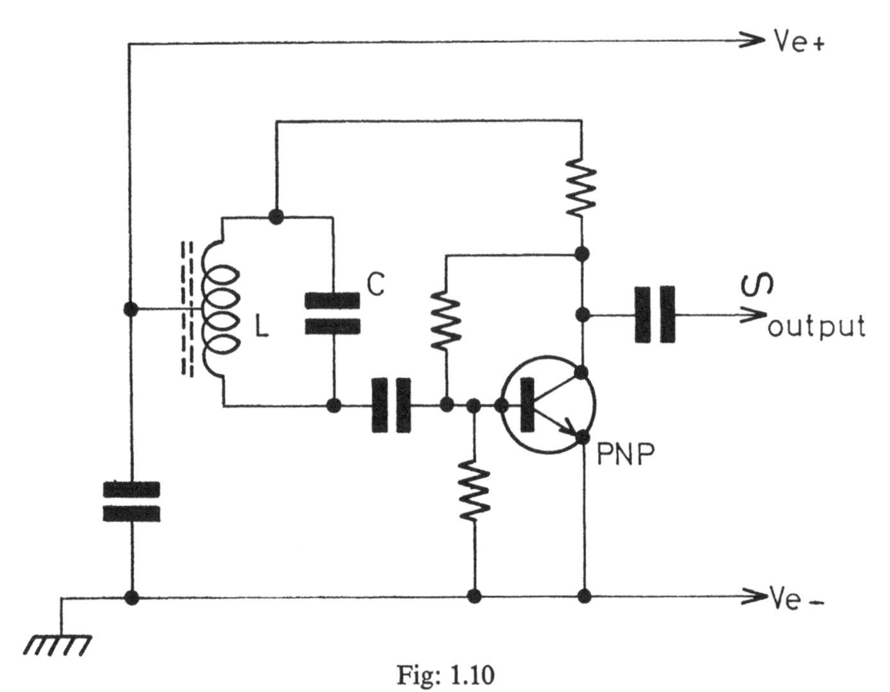

Fig: 1.10
Basic inductive (Hartley) oscillator. Frequency determined by C and L.

is frequently used for electronic organ master oscillators and providing the inductance L is small by comparison with its tuning capacitor C, the frequency stability will be of a high order. This is essential of course for electronic organs which must retain their tuning to concert pitch and for devices like music synthesizers which may also be required to perform in concert with other instruments. There are many different kinds of inductive oscillator circuits but as with all transistorised oscillators precautions must be taken against wide temperature changes and supply voltage fluctuation. Inductive oscillators produce a fairly sinusoidal waveform although distortion can be high due to excessive feedback which may cause the generation of harmonics.

There are of course many types of oscillator that will generate specific waveforms such as square, triangular or sawtooth, exponential and pulse. Specific waveforms are determined by the type of oscillator and the resistive/capacitive or inductive networks used to tune the oscillator and/or shape the waveform. The most common form of square-wave generator is the multi-vibrator shown in Fig: 1.11 although it does not produce a perfectly uniform square-wave. The relaxation oscillator shown in Fig: 1.12A will generate triangular or sawtooth waves, whilst the blocking oscillator in Fig: 1.12B can be designed to produce pulses of very short duration.

The oscillator circuits outlined in the previous paragraphs are but a few of the dozens of different designs possible. What is important to remember is that many of them, like the multi-vibrator, generate a waveform that contains not only the fundamental frequency but a

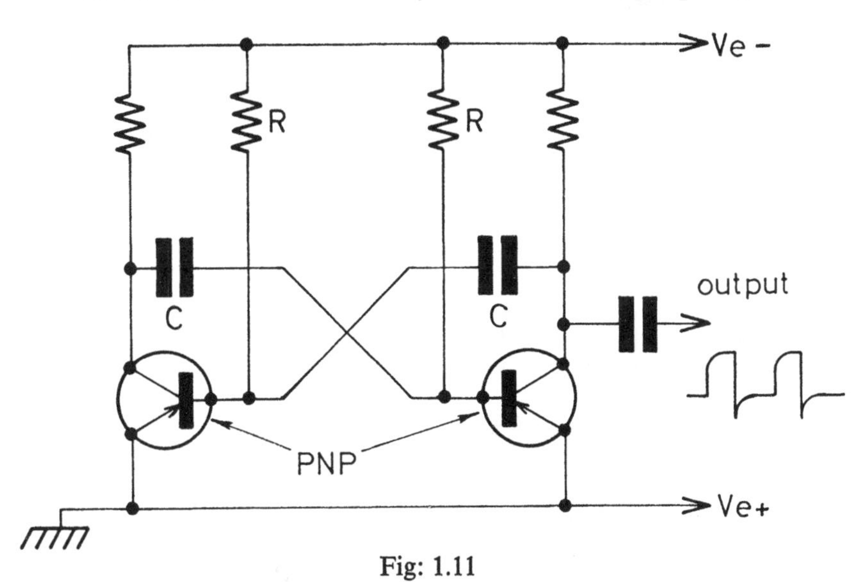

Fig: 1.11
Basic multi-vibrator (square-wave) generator. Frequency and mark-space ratio determined by combinations of C. and R.

whole series of harmonics as well. Even a sine-wave generator can be coupled to a squaring circuit to produce a perfectly uniform square-wave with its attendant harmonics. In order to use these generators as tone sources for electronic organs and music synthesizers, which must be capable of producing specific voices, or indeed the voices of known musical instruments, it is necessary to introduce filters to either extract unwanted harmonics or to re-shape the waveform to that desired. It is not normally practicable to design a tone

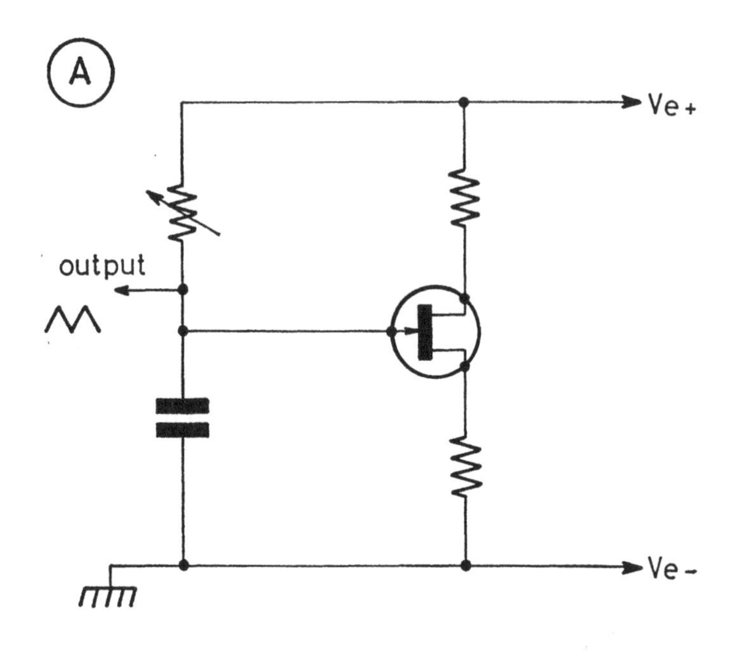
A
Ve+
output
Ve-

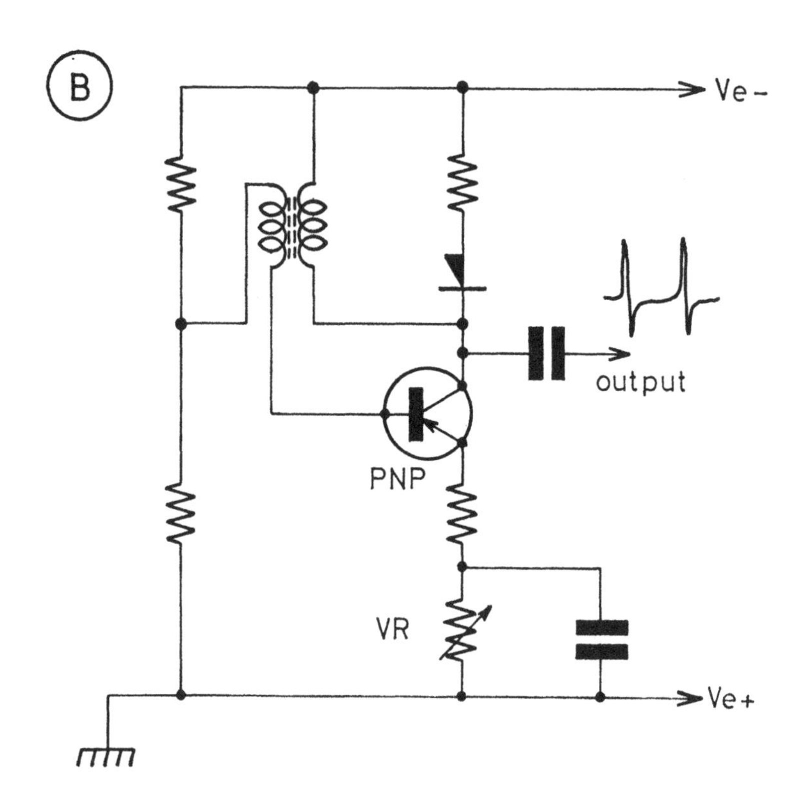
B
Ve-
output
PNP
VR
Ve+

Fig: 2.9 Rotating horn/drum loudspeaker for producing tremulant *(courtesy Thomas Organs Ltd.)*

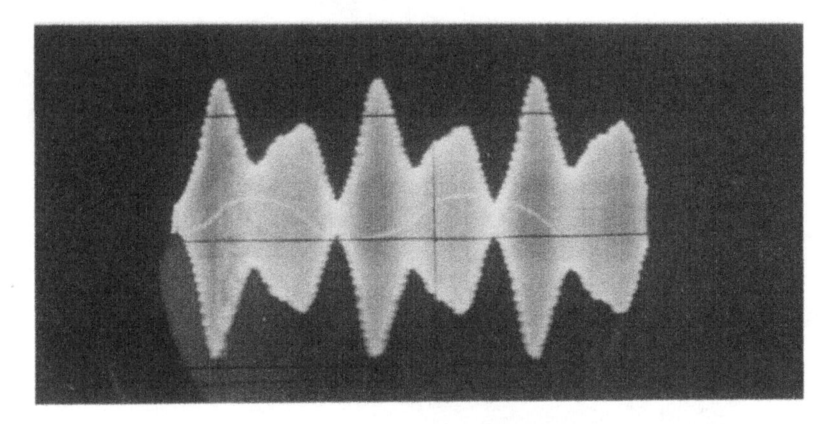

Fig: 2.11 Oscillogram showing the complete waveform, including voice and tremulant, produced by a rotating horn speaker.

Fig: 2.12 Interior of an organ tone cabinet containing a rotating sound deflector (beneath middle loudspeaker) and rotating horn sound transducer (extreme top). Both contribute to an unusual but pleasing tremulant *(photo courtesy Lowery Organs Limited)*.

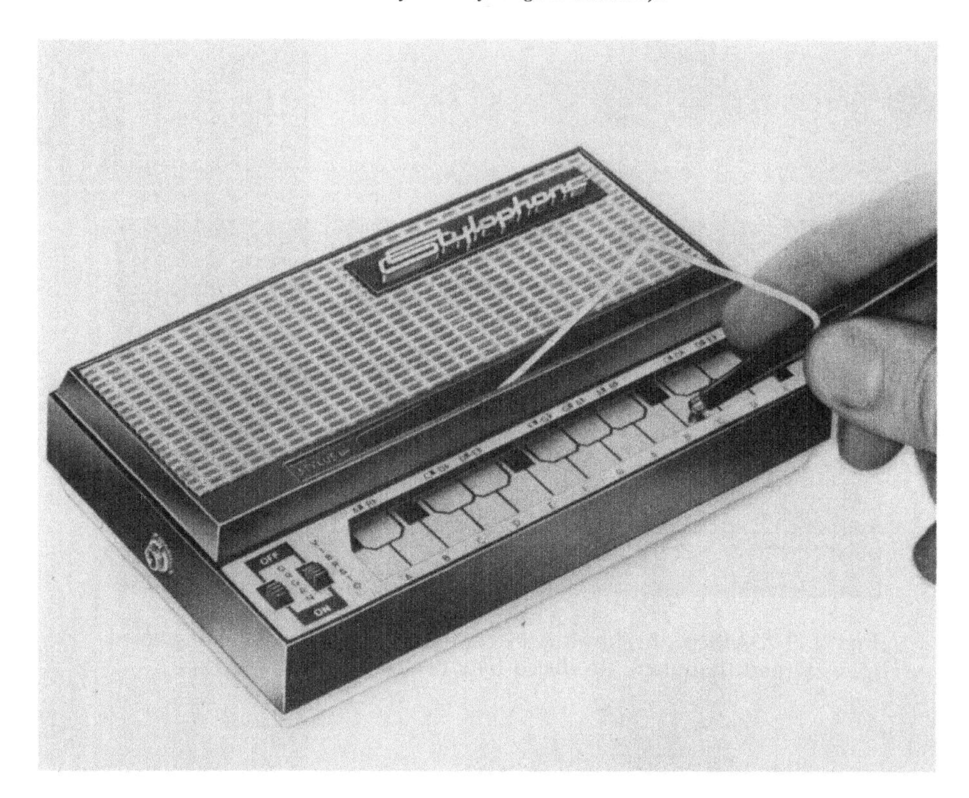

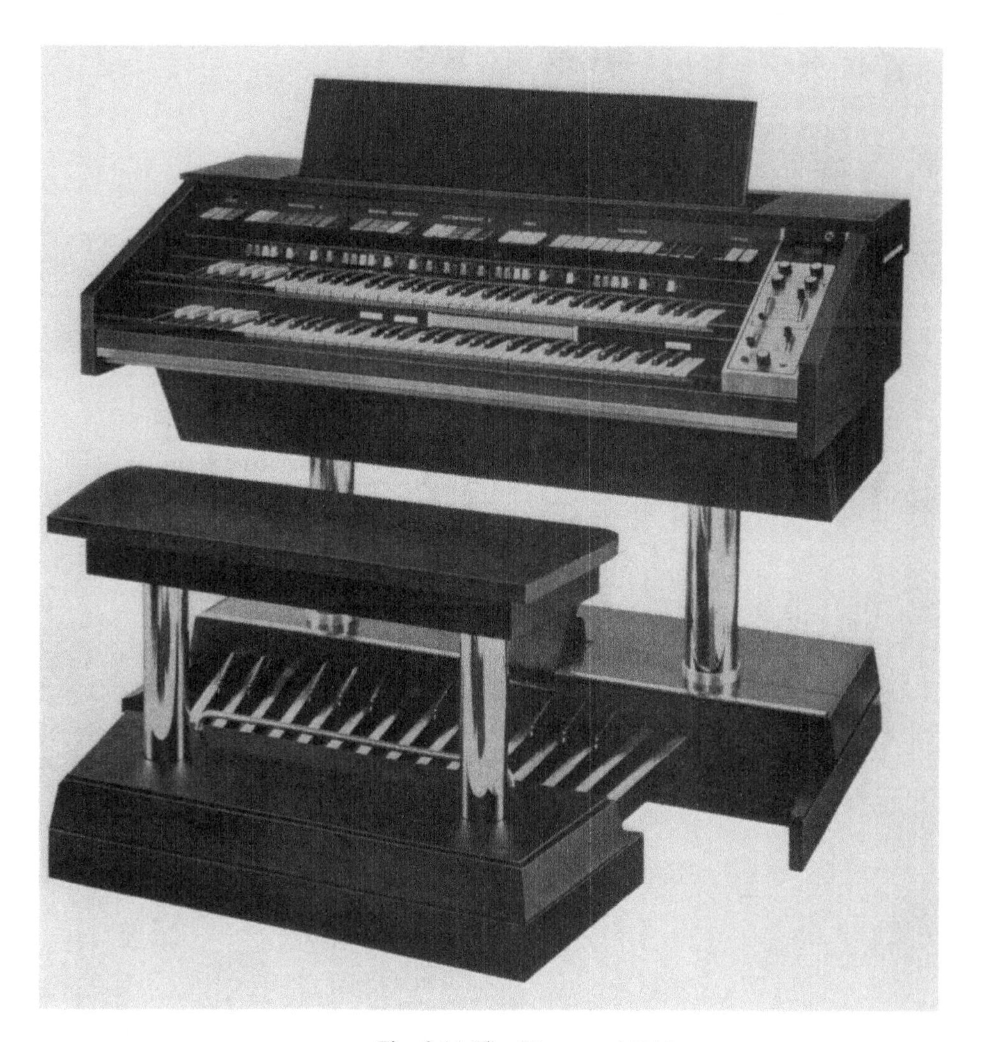

Fig: 2.14 The Hammond X66.

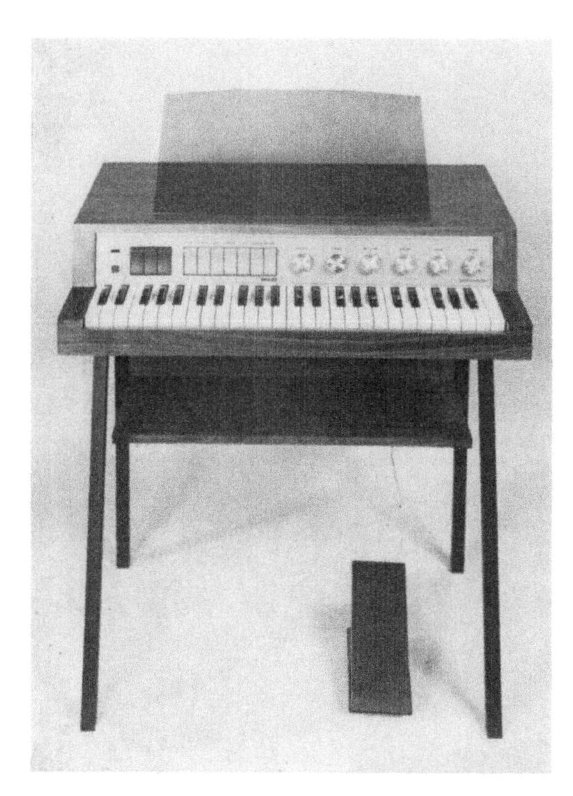

Fig: 2.15 The Philicorda, a small but versatile single manual domestic electronic organ.

Fig: 2.16 The Conn Theaterette. A fairly sophisticated domestic electronic organ with three manuals and percussion – rhythm unit and a built-in tape recorder.

generator with a specific harmonic content, i.e. one that will produce exactly the right waveform for a desired voice. Combining sine-waves of different frequencies to produce a fundamental and a train of harmonics presents the problem of phase locking each component frequency. This can be done with locked tone generators like the Hammond organ tone wheel system which is dealt with in Chapter 2. It is difficult to do with free running oscillators but developments in this direction are in progress. The designer of electronic organs must therefore decide what type of generator to use and then design the voice filters to modify the natural waveform of the generator to comply with that required for the voice. Electronic organs therefore, have several voicing filters for there will be at least one for each voice. Music synthesizers are more sophisticated because they may contain many filters that can be controlled and/or otherwise adjusted to enable the user to select or create whatever voices are required.

Filters

As with tone generators there are an enormous number of possible circuit combinations most of which are derived from basic configurations. Filters used in electronic organs, music synthesizers and the various hybrid electronic musical instruments are normally specially designed for the instruments concerned. However, some understanding of their function may be gained by a study of basic filter circuits which may be composed of inductance, capacity or resistance or combinations of either. The basic function of a filter is to reject or accept one frequency or a band of frequencies as the case may be. Its function is perhaps better understood by comparing it with that of the tuning circuit in a radio receiver which *accepts* the frequency of the station required and *rejects* all others. Equally, a tuning circuit of this kind can be designed to *reject* one frequency and *accept* all others. In this case one could tune out a single station and receive all the rest simultaneously. The tuning circuit in a radio receiver is in effect a filter consisting of inductance and capacity (L

Fig: 1.12 (page 32)

(A) Basic sawtooth oscillator employing a unijunction semi-conductor.
(B) Blocking oscillator to produce pulses of very short duration. Frequency is controlled by VR.

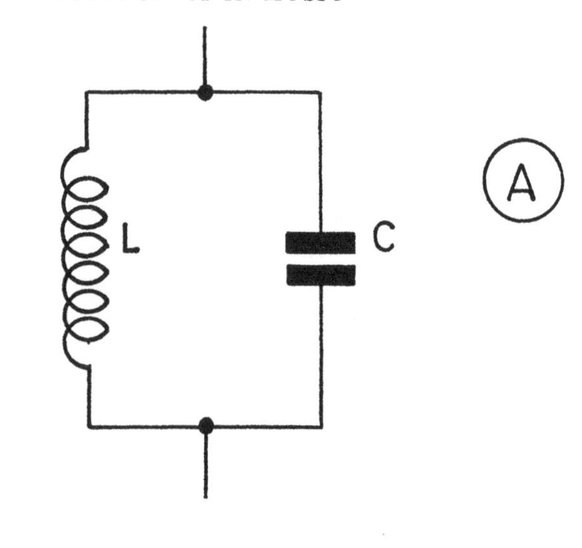

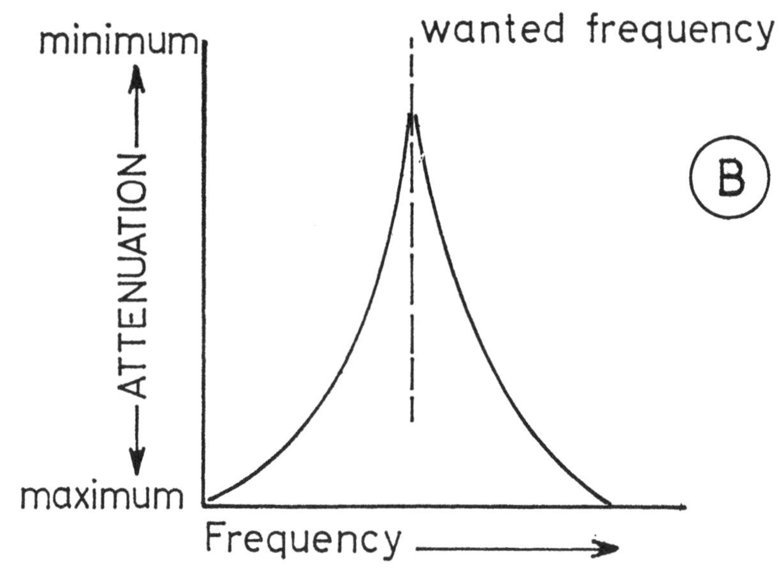

Fig: 1.13
(A) A simple 'tuned circuit' or inductive-capacitive filter. (B) The response of the circuit (A).

and C) as shown in Fig: 1.13A and its ability to tune, i.e. select and pass one frequency and reject all others, is measured by just how much it will reject the unwanted frequencies. A tuned filter of this nature has a high rate of rejection which can be plotted as a graph like that shown in Fig: 1.13B.

Inductive-capacitive filters are normally used when a high rate of rejection or acceptance is required. For example suppose we have a tone containing a large amount of third harmonic and wish to eliminate that harmonic. A filter for doing this would need to be 'tuned' so as to have a high attenuation rate about the frequency of the third harmonic but cause little or no attenuation at the fundamental frequency or at the frequencies of other harmonics if these are to be retained. Filters of this nature can also be designed so that the rate of attenuation can be controlled, e.g. one may require only partial attenuation at a particular frequency. It must be emphasized, however, that inductive filters can seriously distort transient waveforms such as square and sawtooth waves and would not normally be used for tone controlling, for instance, where one is only concerned with gradual attenuation over a band of frequencies such as control over the amount of bass or treble response from an amplifier.

Some fairly typical filter circuits, mostly employing inductance and capacity, are shown in Fig: 1.14 together with their effective frequency responses. The first (A) consists of inductance and capacity in series and unlike the parallel connected circuit in Fig: 1.13 provides maximum *rejection* at one frequency. Fig: 1.14B is a derivation of the series circuit and contains resistance which has the effect of damping the response so as to make rejection or attenuation more gradual. Fig: 1.14C provides gradual attenuation toward the low frequencies or if you like, a gradually increasing acceptance toward the high frequencies.

The combinations of inductance, capacity and resistance as filter circuits are innumerable as also are those composed of only resistance and capacity and which are usually called RC filters. There are, however, three more LC filters that are frequently used in electronics and are simply called low pass, high pass and band pass and, as these names imply, are filters for passing low frequencies, or high

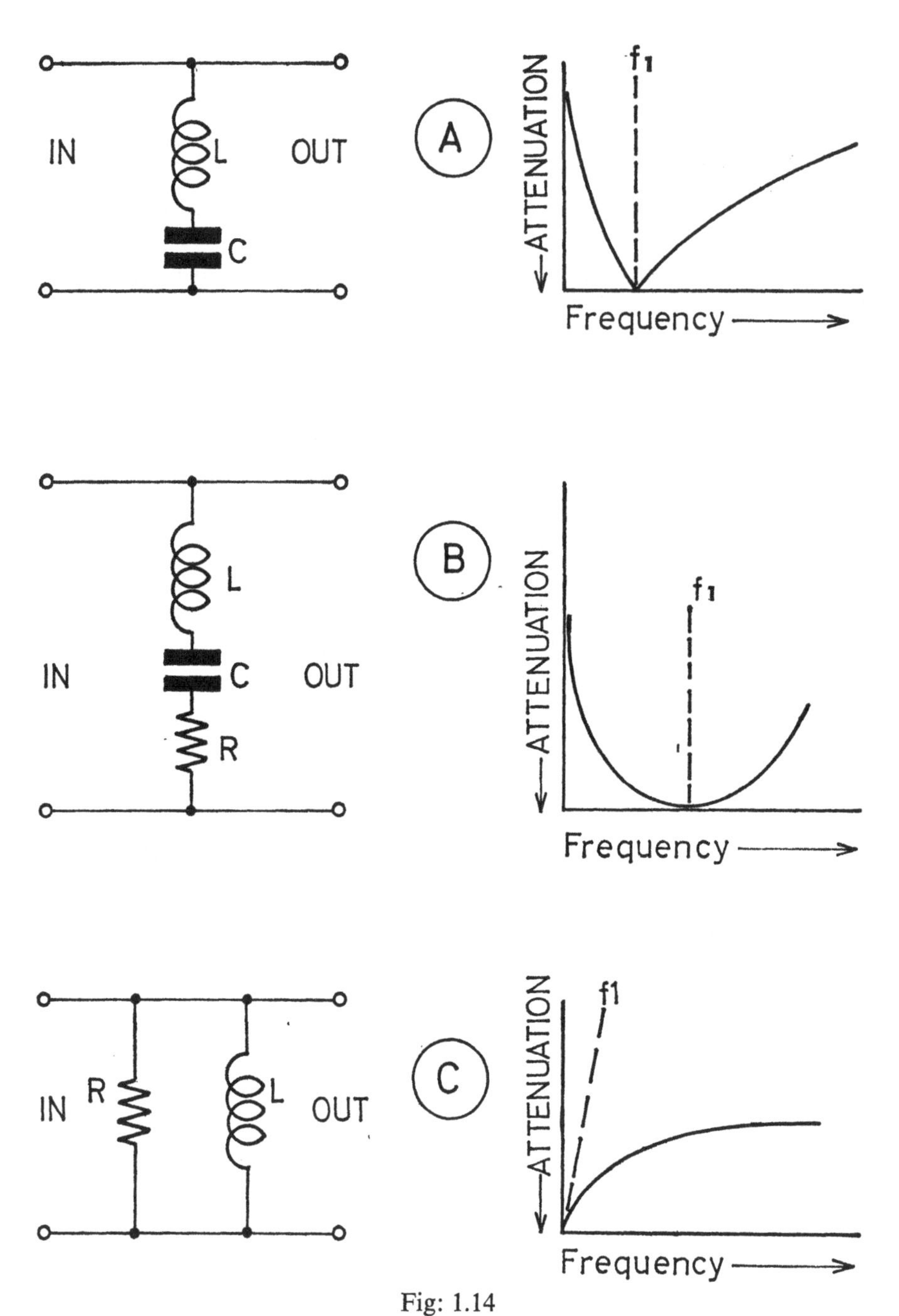

Fig: 1.14
Three typical filter circuits employing inductance capacity and resistance.

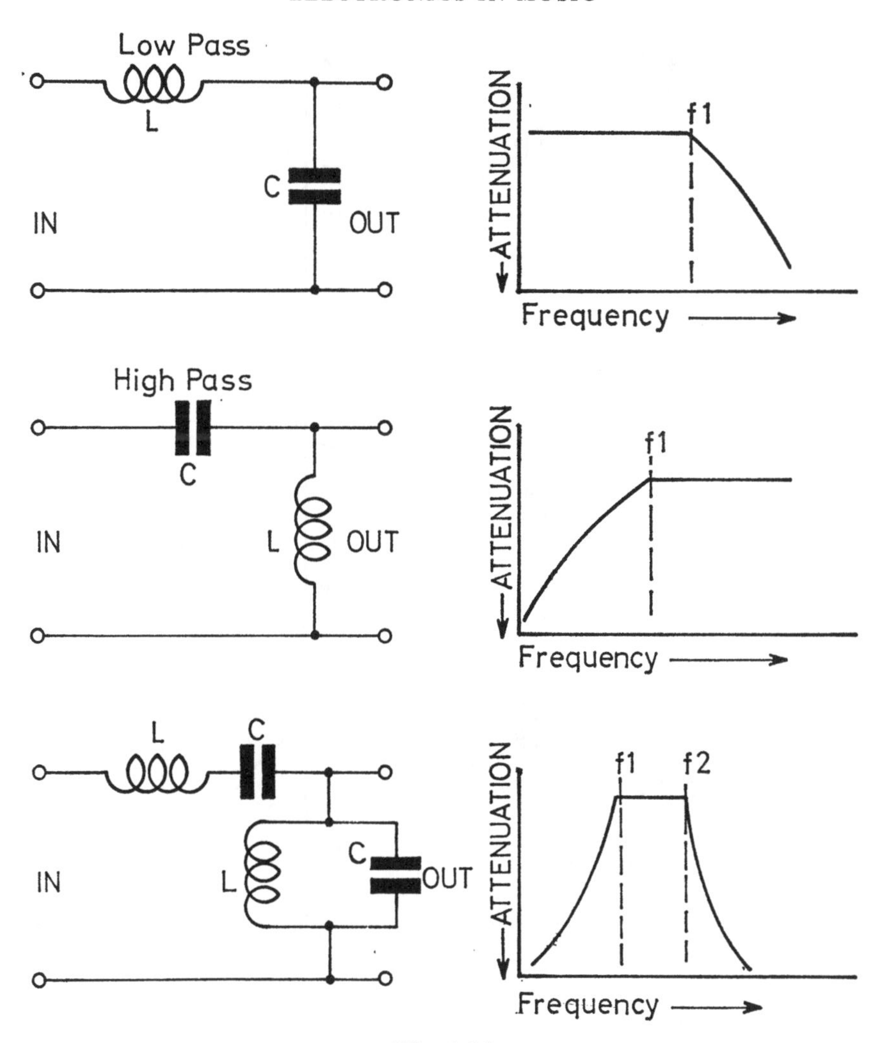

Fig: 1.15
High pass, low pass and bandpass filters and their respective responses.

frequencies, or a band of frequencies. Each is shown in Fig: 1.15 together with its respective frequency response v attentuation graph. Again it must be mentioned that such filters can distort transient waveforms and are not normally used in linear amplifier circuits.

RC filters composed of resistance and capacity are commonly

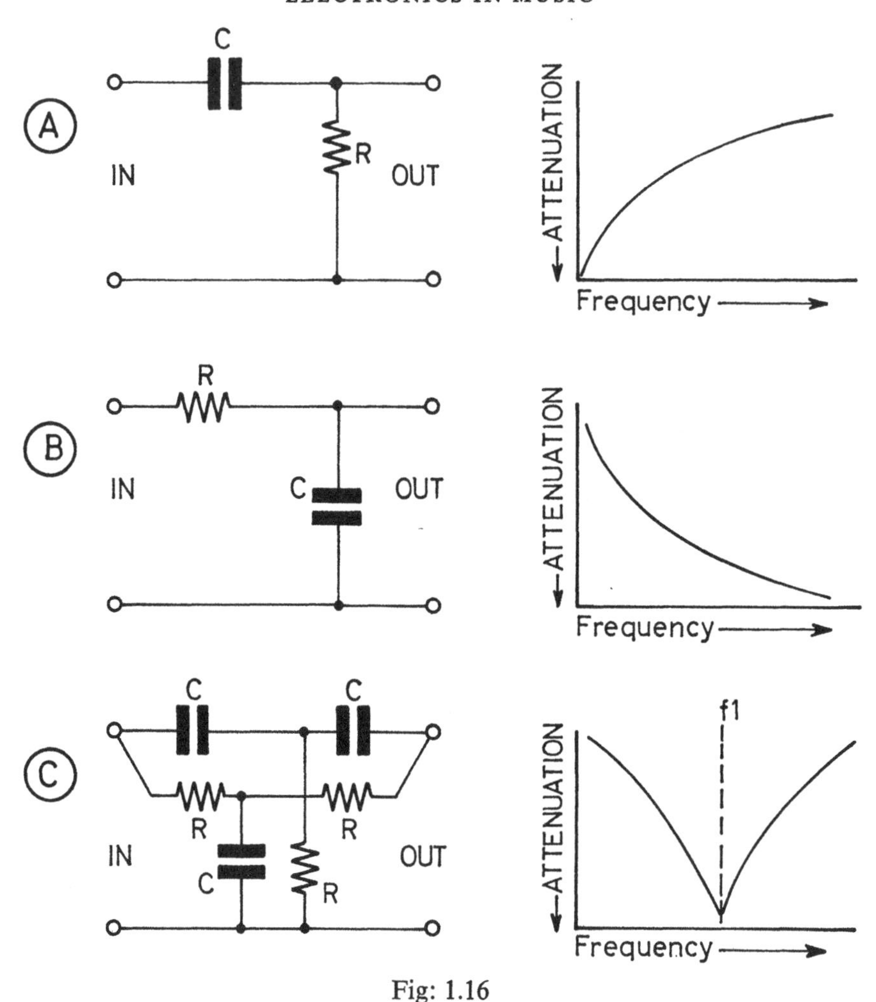

Fig: 1.16
Three typical resistance-capacitance filters and their respective responses.

used in amplifier circuits to avoid distortion of transient waveforms, although in certain cases they are employed for the reshaping of square-waves for example, by differentiation or integration. Certain RC networks can also be used to reject or accept a narrow band of frequencies. Some of the common arrangements are shown in Fig: 1.16 and (A) for example, will provide gradual attentuation toward the lower frequencies. However, to be effective as a filter the values of R and C must be carefully chosen. If C is very large it will have little or no effect, particularly at the very low frequencies. When C is made smaller its reactance to low frequencies becomes higher and in conjunction with R forms a network that provides a degree of attenuation according to frequency. The circuit in Fig: 1.16B is similar except that attenuation increases gradually toward the higher frequencies but this too depends on the value of C, i.e. having a reactance that is high to low frequencies but which becomes progressively lower as frequency is increased. The parallel T network in Fig: 1.16C is an R3 filter for the elimination of one frequency. The values of the capacitive and resistive components can be adjusted for rejection of different frequencies.

Control circuits

It is not sufficient to simply generate a basic tone and then shape the waveform by means of electrical filters in order to produce a musical voice, even though it may not be intended to imitate one that is known. Means of turning the sound on or off for any given period must be provided and the sound may be required to have expression in the form of attack and decay, of frequency variation, of vibrato or tremulant, or may even require the addition of another sound for purposes of realism, particularly in the synthesis of known musical instrument sounds. Any tone generator could be turned on or off by means of a switch or the signal output itself could be interrupted. Alternatively a controlling amplifier could be used, this in turn being provided with a means of cutting it on or off. In this case the tone generator would operate continuously and its output would be coupled to the input of the controlling amplifier. Yet another method of on-off control would be to include a volume control in the output of the generator and simply turn this up or down to interrupt the signal. If a switch, or better still a fast acting contact, is used to interrupt the signal then the starting and stopping would be almost instantaneous. With a volume control this would be progressive, i.e. the signal would increase gradually as the control was turned up and decrease gradually as it was turned down. With a combination of switch and volume control one could obtain an instantaneous 'ON' and a gradual 'OFF'. But this is too cumbersome especially if a number of different tones are to be sounded in fairly rapid succession. Instant ON or OFF or instant ON and gradual OFF can be done much more easily by electronic methods even though a switch or a keyed contact may be necessary to trigger off the operation. A typical example is the sustain effect that can be produced by most electronic organs. The tone generators are fed to an electronic control circuit which when keyed, instantly turns the tone ON and then allows it to slowly turn OFF, or decay. The principle can be extended to provide a slow ON (soft attack) and a fast decay, or in fact to provide any degree of attack and decay.

The electronic methods for doing this are numerous indeed and range from a simple contact that merely allows a signal to pass when opened or closed but which does not provide any specific form of attack or decay, to the complex circuitry employed in music synthesizers which can be used to generate an almost infinitely variable degree of attack and decay. Despite their complexity, however, almost all circuits used for sustain and percussion effects in electronic organs and for controlled attack and decay in music synthesizers, necessitate the use of *timing circuits*, i.e. a combination of capacity and resistance for determining the rate at which the

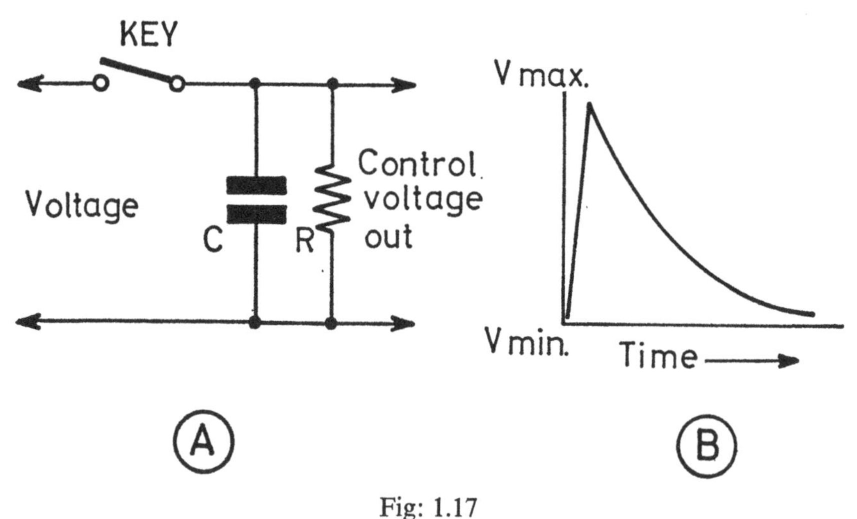

Fig: 1.17
Basic method of obtaining a control voltage with a fast rise time and slow decay.

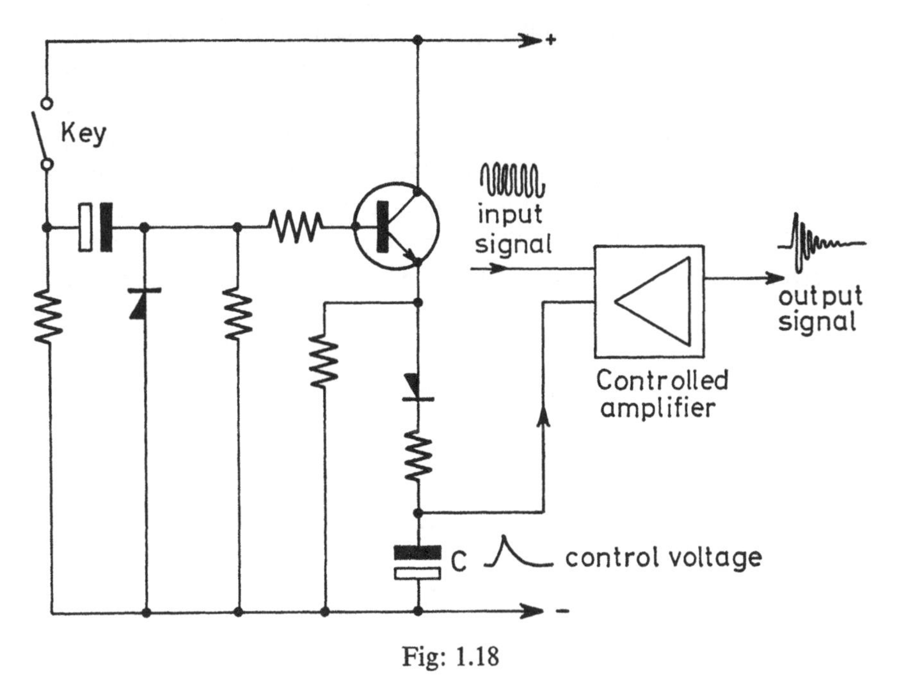

Fig: 1.18
Basic circuit for obtaining a control voltage for attack and decay.

control voltage rises (attack) and/or the rate at which it decays. The circuit given in Fig: 1.17A is a typical example and in this a voltage is used to first charge the capacitor C. This occurs almost instantaneously (fast attack) when the key is closed. The mechanical requirements of the system are that after being closed the key would instantly open again. This leaves the capacitor charged whereupon it commences to discharge gradually through the resistance R. The result is the production of a control voltage that quickly rises to maximum (V max) and then slowly falls to minimum (V min) over a period of time as in Fig: 1.17B.

One of many possible circuits incorporating an RC timing control is shown in Fig: 1.18. This circuit employs a transistor which is instantly switched on when the key is closed. This produces a large positive pulse at the emitter which in turn, charges the capacitor C. The load imposed by the amplifier being controlled behaves as a resistance which gradually discharges C. This is in fact a simple form of voltage control, not unlike that employed in music synthesizers.

Another method, sometimes used in electronic organs, is one which employs a light dependent resistor (LDR) and which provides a more linear form of control than can be obtained with circuits like that shown in Fig: 1.18. The LDR is a device which changes its self resistance when light shines upon it. In the dark the resistance is quite high but when the light sensitive element is illuminated the resistance falls to a very low level. The resistance value therefore,

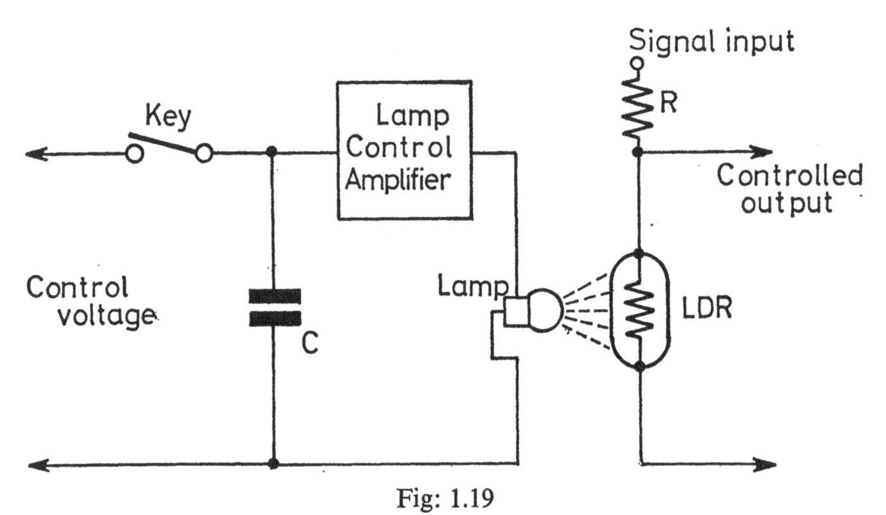

Fig: 1.19
Use of a light dependent resistor (LDR) for signal control.

varies with the intensity of the illumination. The LDR can be used as part of a potential divider and its action is quite fast. Instantaneous illumination produces an almost instantaneous fall in resistance. It can, therefore, be used to switch a tone ON by instant illumination and then, providing the light is gradually extinguished, the tone will be gradually switched OFF, i.e. its level will slowly fall or decay. In the basic circuit shown in Fig: 1.19 the brilliance of the lamp is controlled by an amplifier. When the capacitor C is charged the amplifier is cut off, the lamp is extinguished, the LDR is at high resistance and the input signal is allowed to pass. As C slowly discharges the amplifier conducts, the brilliance of the lamp is gradually increased and the resistance of the LDR slowly decreased, which in turn, causes a decay or gradual diminution of the signal. It should be emphasized that this is only one way of employing an LDR. There are many variations of the idea.

The simple methods of control outlined in the previous paragraphs do not necessarily represent the techniques likely to be employed in modern electronic organs and music synthesizers, but merely illustrate one or two basic principles. Control in synthesizers for example can involve the use of literally hundreds of semiconductors and components in circuits far too complex to describe in detail in this book.

Vibrato and Tremulant

It was explained earlier that vibrato is generally accepted as a pitch variation which is rhythmic, i.e. changing at an even rate and that tremulant is a loudness variation, also rhythmic, with fluctuation at an even rate. Both these effects can be produced electronically and the rate of vibrato or tremulant is usually in the region of 5 to 10 fluctuations per second. In electronic organs both effects may be available but with electric guitars it is only possible to produce a tremulant, i.e. a loudless fluctuation electronically. Guitar vibrato or pitch variation is produced mechanically. In the case of plectrum guitars by mechanically altering the length of the strings by a very small amount and in the case of Hawaiian or pedal steel guitars by manipulation of the playing steel.

The basic electronic method of producing vibrato is to use a sine-wave oscillator running at between 5 and 10Hz per second (control is usually variable) and then apply the sine-wave voltage to the tone generator. The phase shift oscillator (Fig: 1.9) is frequently used not only for vibrato control but for tremulant as well. For vibrato the frequency of the sine-wave voltage determines the frequency of the vibrato and its amplitude determines the depth of the vibrato, i.e. amount by which the frequency of the generator is changed. For tremulant the amplitude of the control oscillator determines the amount of variation in loudness. The diagrams in Fig: 1.20 may help to explain. That shown as (A) illustrates vibrato and (B) the effect of tremulant.

There remains now the tremolo effect often produced by players

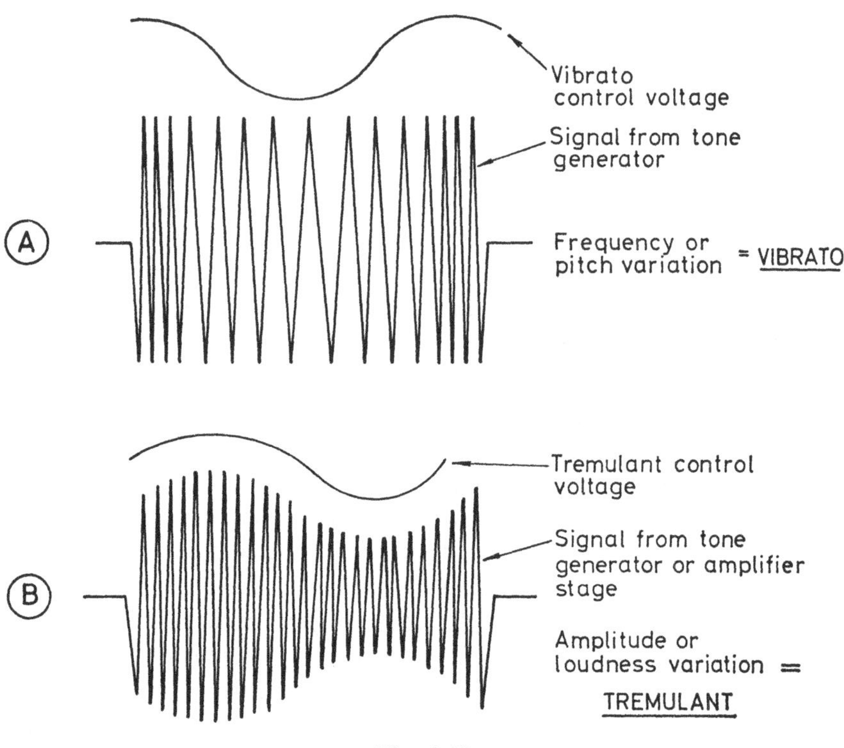

Fig: 1.20

of stringed instruments like the mandolin, which is a rapid repetition of a single note. The glissando effect of starting a note at one pitch and then gradually or rapidly gliding to another pitch can also be produced by electronic means. The rapid ON-OFF sounding of a note is achieved by applying a rapid ON-OFF control voltage to the tone generator or to an amplifier carrying the tone. A multi-vibrator like that shown in Fig: 1.11 could be used as this circuit can operate at slow speed but generate the requisite fast ON and fast OFF control voltage. The diagram in Fig: 1.21 may serve to illustrate the principle.

Glissando requires the application of a slowly rising (or falling) control voltage applied in such a way as to gradually bring a tone generator up or down to another pitch i.e., to one different from the starting pitch. There are numerous ways in which this can be accom-

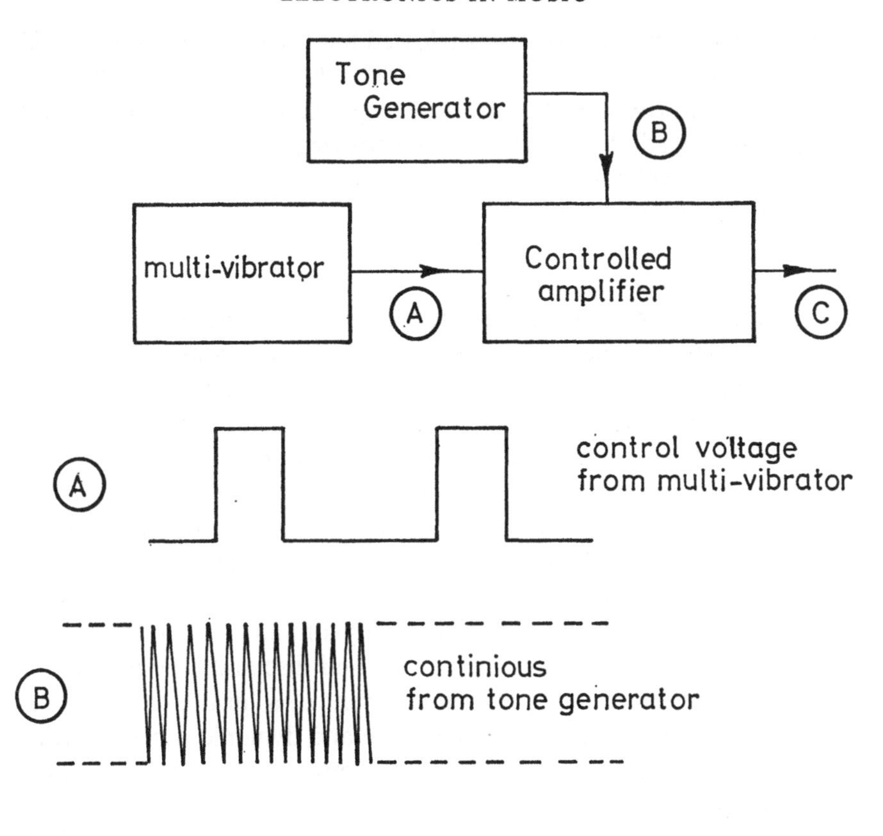

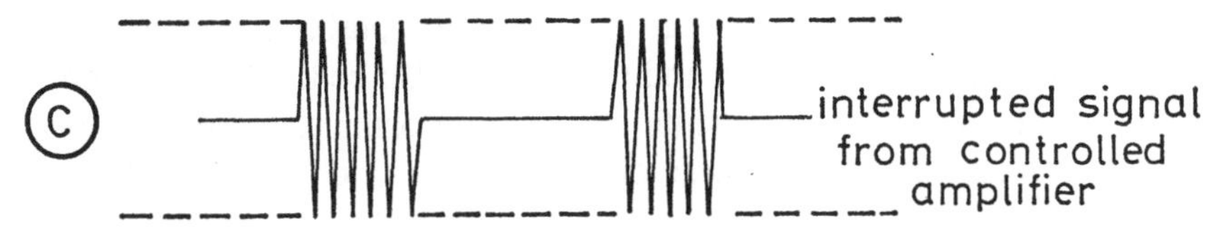

Fig: 1.21

plished electronically and one method used in electronic organs is to apply an initial pitch setting voltage to all the tone generators simultaneously to bring them down by an interval of about one whole tone, for example from C to B. This is done by means of a foot operated switch attached to the side of the swell pedal. On playing the note, or notes, the switch is released but the voltage it has applied to the tone generator falls slowly and allows the pitch of the tone generators to rise i.e., to return to true pitch. This particular method was and may still be used by the Lowery Organ Company. The basic principle of voltage controlling the tone generators is commonly used in music synthesizers.

Sound reproducing systems

Sounds and music generated electronically can only be reproduced for listening by means of amplifiers and loudspeakers. Without these, the electronic organ for instance would be a dead instrument. Having gone to a great deal of trouble to generate the signals that will ultimately become audible sounds it is important that they are not distorted or changed in any way whatsoever when they are amplified to sufficient level for audible reproduction. This applies equally to the recording of the signals on magnetic tape or disc, etc., for reproduction at a later time.

The nature of the domestic high fidelity amplifier and loudspeakers for reproduction of radio programmes and disc and tape recordings is fairly well understood. The requirement of a 'hi-fi' system is that it must have a wide frequency response and dynamic range in order to faithfully reproduce not only the lowest but also the highest frequencies, especially those at which harmonics occur, and produce as closely as possible the relative loudness variations of the original material be it music or sounds. Distortion caused by the system as a whole must not be greater than a certain acceptable maximum and spurious sound such as hum and noise must also be limited to acceptable levels. These requirements are not, unfortunately, always met by manufacturers of electronic organs and electric guitar amplifiers. It is not uncommon to find a high degree of distortion and hum and noise being produced by the final amplifier stages in electronic organs and by electric guitar amplifiers. The sounds from a music synthesizer could be completely spoiled by a poor recording or reproducing system. Sounds and music generated electronically should always be reproduced by the highest quality amplifier and loudspeaker systems and/or recorded, whether on tape or disc, by the best possible recording equipment.

Technology

It is impossible to avoid the use of technological terms in books dealing with technical subjects and yet some of the terms already used in this book could entail yet another book to explain them in detail. References to various useful textbooks devoted to specific subjects will be found in the Bibliography at the end of this book.

CHAPTER TWO

Electronic Musical Instruments

Electronic musical instruments were only a dreamed of possibility until (*circa* 1907) the scientist Lee de-Forest put the 'grid' into a diode valve and produced the thermionic triode valve which was capable not only of generating electrical signals suitable for musical voices but of amplifying them as well. This immediately made it possible to design musical instruments with voices generated and simulated entirely electrically and capable of sound power equal to, or greater than could be produced by conventional instruments. Even the enormous power of the church organ could be equalled. Another possibility was amplification of the sound of any conventional musical instrument by means of a microphone, amplifier and loudspeaker and even of 'electrifying' existing instruments, particularly those employing steel strings, like the piano and guitar. The years following the end of the 1914–18 war saw considerable development in electronic and electrified or electrically amplified musical instruments. Domestic electronic organs evolved, the electric guitar became an almost standard solo instrument in popular dance music orchestras and giant electrically controlled theatre organs entertained the patrons during the intervals between films at the cinema and became popular on the radio. Since the 1939–45 war the electronic organ has almost replaced the piano as a family 'front room' musical instrument and the electric guitar in its various forms is now the 'orchestral' instrument for practically all 'pop' music. There have of course been many hybrids like the Bechstein electric piano, electric violins and 'cellos and the Theremin, etc., and more recent developments in electronics have produced electronic pianos with various 'extra' voices, i.e. honky tonk or tinny piano, harpsichord, etc., as well as electronic organs with versatility bordering on that of the music synthesizer. Indeed some electronic organs now include a small built-in keyboard synthesizer.

The Electronic Organ

These days electronics are pretty much taken for granted and few know or even care much about the complex assemblies of tiny components and transistors that go to make household devices like TV sets, record players and tape recorders, etc. Apart from electro-mechanical components, such as loudspeakers, the modern electronic organ too consists of many separate circuits each designed to do its own special job. When talking of these circuits collectively, engineers generally refer to them as 'the electronics'.

Nearly all electronic organs now employ transistors and the 'printed circuit board' technique for the assembly of individual or collective circuits. For example, a single note or tone generator may simply consist of the requisite transistors and components for the circuit which are assembled on a small board of insulating material such as paxolin and connected together with wires that look like thin copper strips stuck onto the board. One of the advantages of printed circuit technique is that if a fault occurs in an individual circuit board the board itself can be quickly changed for another, thus saving an engineer the time taken to trace the fault.

Basic principles (see Fig: 2.1)

On most electronic organs the top can be lifted to gain access to the many printed circuit boards each dotted with components and all connected together and to keying contact rails and voice tabs, etc., by seemingly complex bunches of coloured wires. There are, however, a number of mechanical parts in an electronic organ and of these are the keyboards and pedal keys. Each key or pedal may actuate one or more electrical contacts, these in turn allowing the notes to sound via the loudspeakers. The electronic tone generators are the basis of the electronic organ and there are normally twelve, each one being tuned so as to cover one complete octave in chromatic tempered scale. The 'pitch' of the whole octave may be at the highest the organ keyboard will allow, or even one octave higher i.e., beyond

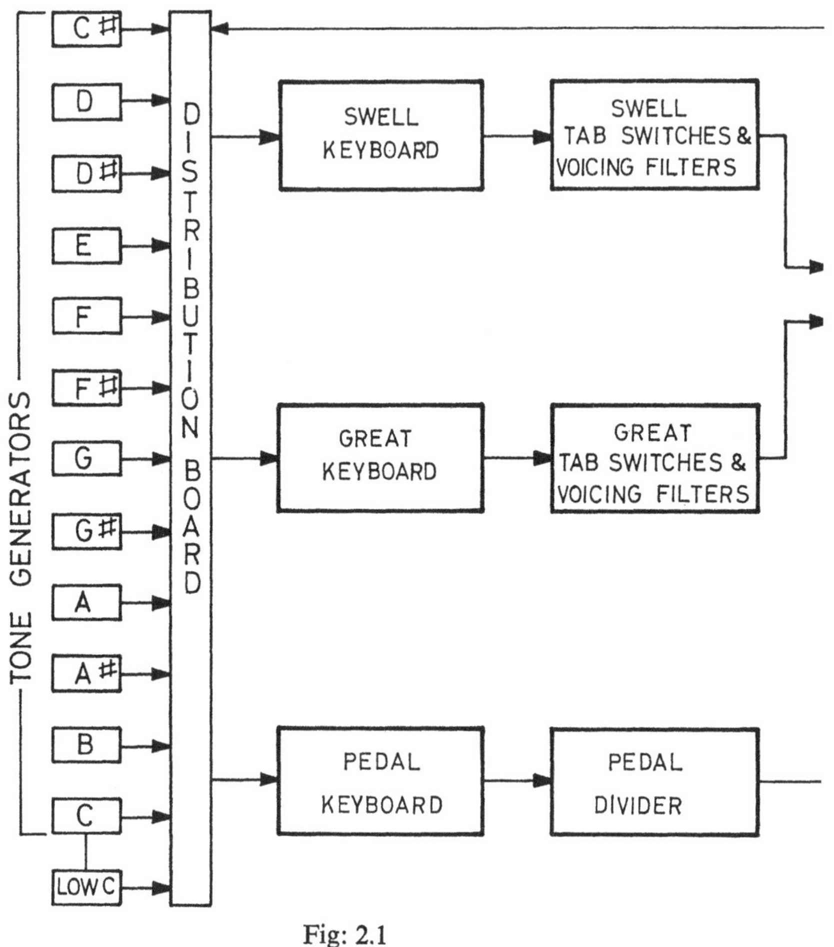

Fig: 2.1
Schematic showing the sections of electronics that go to make an electronic organ. (*courtesy Thomas Organs Limited.*)

the normal upper range of the keyboard. All notes lower than those set by the tone generators are obtained by 'dividers' and enough of these are included to obtain all the notes right down to the lowest on the keyboards and also the very low pitched pedal notes. In most organs each tone or note generator and its dividers may be assembled on a single circuit board.

Before the tone signals can be sounded, however, they must be

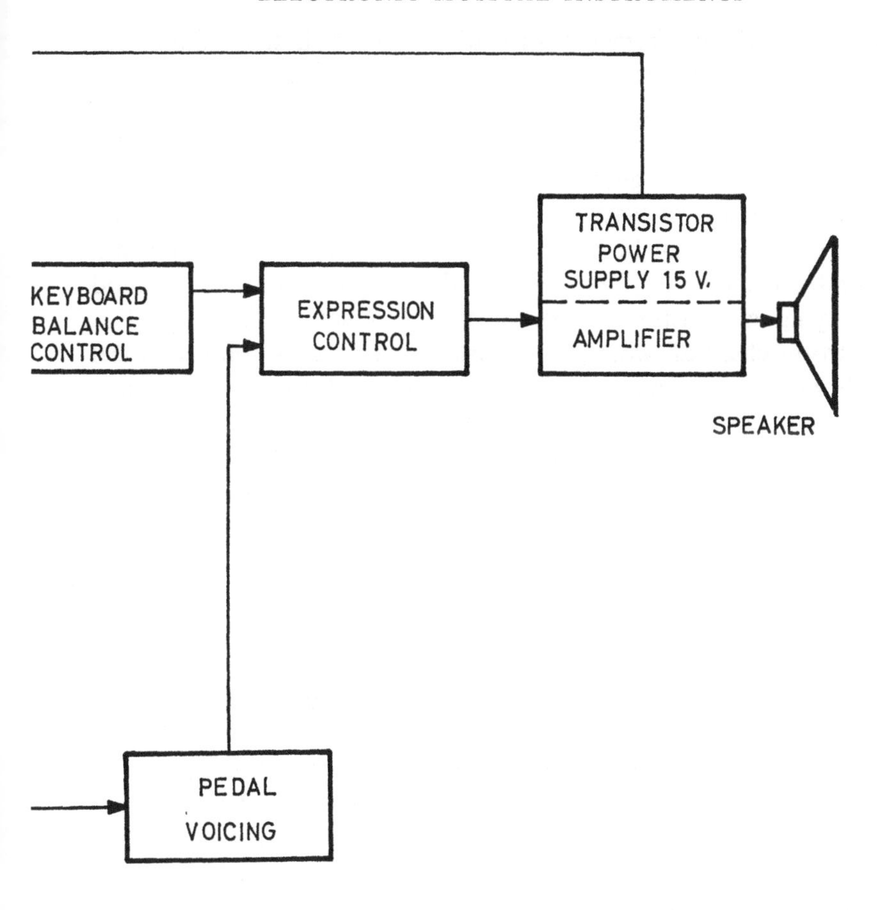

'voiced', i.e. given their characteristic sound. Techniques similar to those outlined in Chapter 1 are employed for this and also for special effects such as sustain and percussion, etc. Only after voicing and other special treatment such as that required for sustain, etc., are the notes allowed to pass to the swell or expression pedal and from there to the main amplifier and loudspeaker(s). The swell pedal is simply a volume control operated by a foot pedal and which, on most domestic organs, simultaneously controls the volume of sound from the whole organ, i.e. the manuals and pedal bass. Some organs have separate pre-set volume controls, one for each manual and pedal bass respectively, so that loudness can be balanced to the player's requirements. In addition to the circuits required for tone generators and voicing, etc., there will also be at least one for electronic vibrato and/or tremulant effects.

Electronic organ tone generators

There are several ways in which the fundamental notes for an electronic organ can be generated, but which can be resolved into three basic methods. First there is the purely 'electronic' method employing a transistor, valve or neon oscillator. The second is 'electro-mechanical' which may consist of a rotating toothed wheel made of magnetic material (iron or steel) and an electro-magnet. This method is well known to Hammond organists as the 'tone wheel system'. Thirdly there is the 'mechanical-optical' method that utilizes a rotating transparent disc on which the tone 'waveforms' are drawn and are scanned by a light sensitive device. There is a fourth method that might be described as 'electrical-mechanical' and was used at one time in an organ known as the 'Electrone' made by the John Compton Organ Company. This generator employed a fixed disc on which the waveforms were engraved and then scanned 'electro-statically'. There have been many variations based on these methods but the two most popular today are the 'electronic' and 'electro-magnetic', i.e. the transistor oscillator and the tone wheel generator. Most modern domestic electronic organs, however, employ transitors not only for tone generators but for all the complex circuits necessary for voicing and effects and of course the amplifiers.

The transistor tone generator or oscillator circuit must above all be stable, i.e. it must not wander off tune. For this reason most transistor tone generators are inductive which means they employ a tuned circuit consisting of a coil with a core of compressed iron dust or ferrite material and some parallel capacitance. The core can be screwed into or out of the coil and is invariably used for fine tuning. A circuit of a typical transistor inductive oscillator is shown in Fig: 2.2. At one time the multi-vibrator circuit (Chapter 1.11) was used but unless special precautions are taken the multi-vibrator has a tendency to drift off tune and is now rarely used. However, a specially modified version of this circuit is very widely used as a frequency divider.

Normally there are 12 electronic tone generators in an electronic organ and these are tuned to produce a chromatic tempered scale from C and going down to B. The next C below B is produced, as all other lower notes of C, by dividers. Normally the actual pitch of the tone generators will be that of the highest octave of the upper manual. A few organ manufacturers do, however, pitch the tone generators one octave higher, i.e. one octave above the highest keyboard octave in which case *all* the keyed notes are derived from dividers.

The Tone Wheel System

Electro-magnetic tone generators known as 'tone wheels' are used in a system exclusive to the Hammond Organ Company although a similar system was in fact employed in the Robb Wave Organ made in Canada in the 30's by Morse Robb of Montreal. The basic Hammond tone wheel generator is shown in Fig: 2.3. A toothed wheel made of magnetic material, e.g. soft iron and is rotated close to an electro-magnet so that each tooth generates a small electric current in the coil of the electro-magnet, as it passes. If, for example,

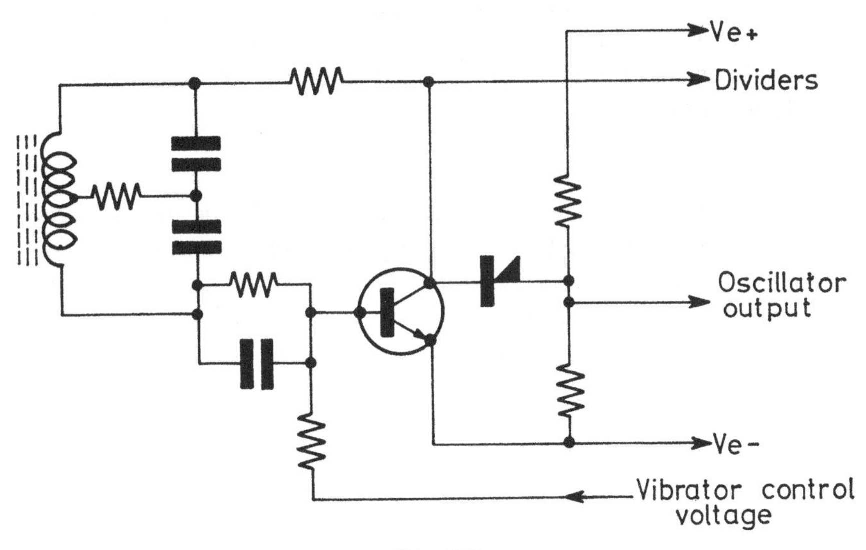

Fig: 2.2
Typical organ tone generator circuit. (*courtesy Thomas Organs Limited.*)

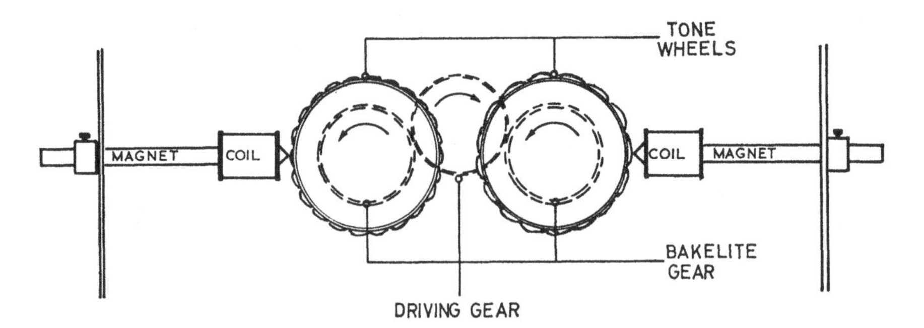

Fig: 2.3
The Hammond Tone wheel method of tone generation. (*courtesy Hammond Organs Limited.*)

the wheel had 440 teeth then 440 pulses of current would be generated for one complete revolution per second of the wheel. This would produce the fundamental pitch of A at a frequency of 440Hz. In the system up to 90 tone wheels, each with a different number of teeth, are used and are simultaneously driven by a constant speed motor which makes it virtually impossible for the system to go out of tune. Not only are the fundamentals derived from the system but also all the harmonics as well.

Dividers

In electronic organs which employ purely electronic tone generators it is not very practicable to employ a separate generator for every note of the keyboard and pedal bass. The alternative and most common arrangement, as already mentioned, is to use 12 generators and from these produce all the other notes, right down to those required for the pedal bass, by means of frequency dividers.

A frequency divider is simply an electronic circuit that produces one cycle of oscillation for every two cycles fed into it. For example, let's take a fundamental note, as produced by a note generator, of 1046Hz (cycles) which is approximately the pitch of the note C two octaves above middle C. If this pitch (1046Hz) is divided by 2, the next octave down will be obtained which is C at 523Hz (approx.).

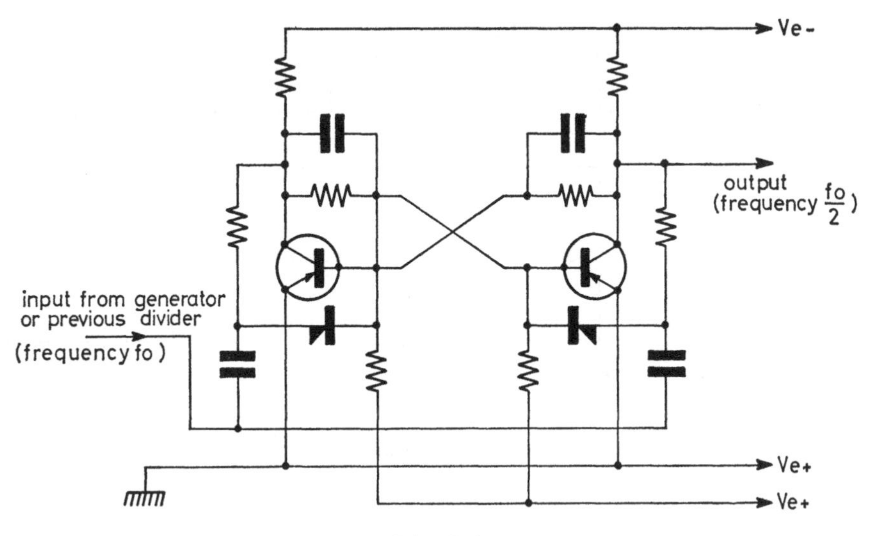

Fig: 2.4
The Eccles-Jordan frequency divider circuit. Many derivations of this are used as frequency dividers in electronic organs.

By dividing again, i.e. by dividing C at 523Hz, the note of C 261Hz (approx.) or middle C will be produced and we can in fact go on dividing to obtain the lowest notes for the keyboard as well as those for the pedal notes which may include C at 130Hz (approx.), C at 65Hz (approx.) or even the lowest used C at a frequency of approximately 32Hz.

The Bi-Stable divider

One of the most common circuits used for frequency dividing, although there are many derivations of it, is the Eccles Jordan bi-stable arrangement shown in Fig: 2.4. The function of the circuit is such that one transistor produces one half-cycle of oscillation for every complete cycle fed into it. Both transistors, therefore, produce a complete cycle from every two cycles of the driving signal. In other words the circuit divides the input frequency by two. If the output from one divider is coupled to another, the frequency can be divided yet again. The diagrams in Fig: 2.5 may serve to clarify this. The top wave-form (a) shows several cycles of a note generator. If

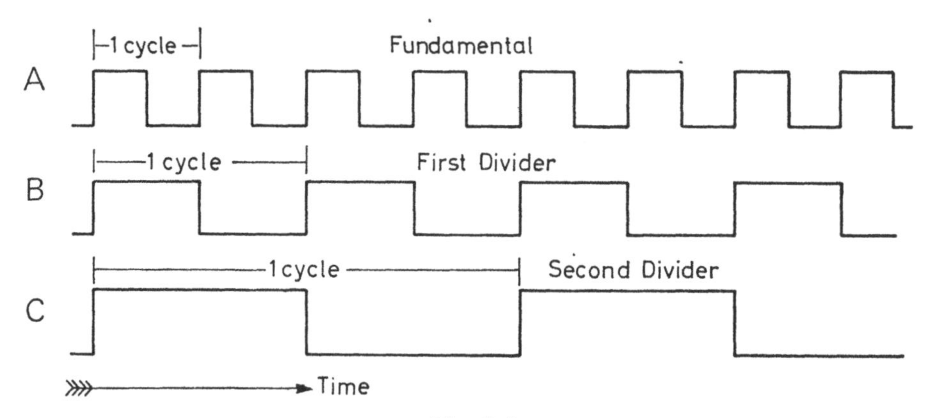

Fig: 2.5
The principle of frequency dividing.

these are fed to a divider the frequency will be halved as shown by wave-form (B) and, if divided again, will appear as waveform (C). Square-wave forms have been used in the illustration because the ouput from a typical divider is a square-wave. The wave-form produced by a fundamental note generator may not, however, be square in which case it would have to be squared or modified in such a way as to become a pulse of very short time duration. A signal with a fast leading edge is necessary for driving a frequency divider. One of the main advantages of the divider system is that the outputs of all the dividers will be 'phase locked' with the master tone generator although should an oscillator go out of tune then the dividers will go out of tune with it.

The block diagram given in Fig: 2.6 shows a typical note generator and divider system. In this case the chain is for all the notes of C from the highest to the lowest required by both the keyboard manuals and the bass pedals. There would of course be twelve separate divider chains, one for each note of the chromatic scale. In the example the chain begins with the fundamental note generator pitched at the highest C or at 1046.502Hz. The first divider therefore, produces C at 523.251Hz, the second C at 261.625Hz (middle C) and the third C at 130.812Hz. On some organs the keyboard range might be extended by raising the pitch of the note generators by one

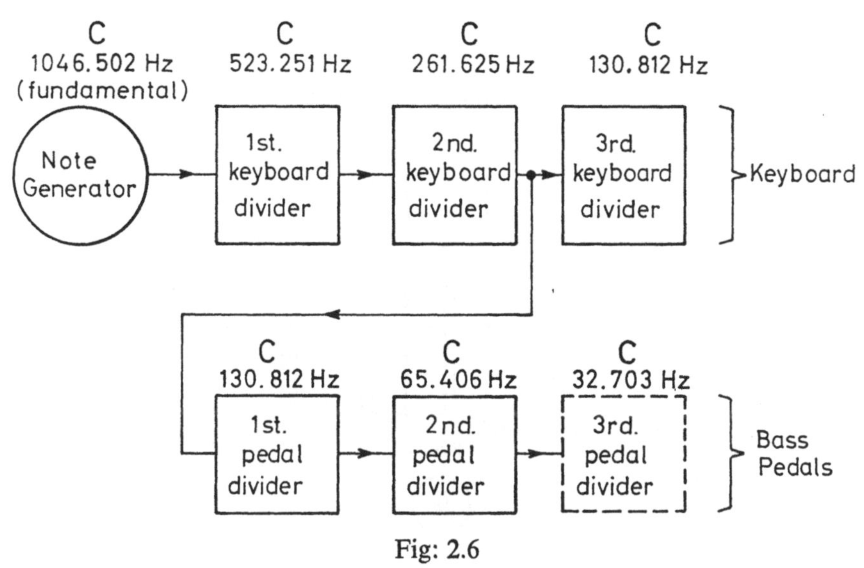

Fig: 2.6
Typical tone generator – frequency divider chain used in electronic organs.

octave and the addition of another set of dividers. In the example given the note generator would be raised to C at 2093.004Hz. Extension in the other direction could be obtained simply by adding more dividers. It is usual to provide a separate divider chain for the pedal notes and this may be tapped off the keyboard chain as shown. The last divider in the pedal chain (C at 32.703Hz) would normally only be used on large organs with a full 30 or 32 note pedal range. The keying of individual notes takes place after dividing and this is followed by tone formation or voicing and the routing of individual registrations and voices, etc., via their various selector tabs.

Voicing

All acoustic musical instruments, including the pipe organ, have a fundamental pitch range of course but directly generate their own harmonics i.e., their voices are self produced and depend largely on the physical construction of the instrument and its method of producing sound. The piano for example always has its own characteristic timbre and cannot be switched to sound like a violin or a clarinet. The tone generators in an electronic organ do not directly

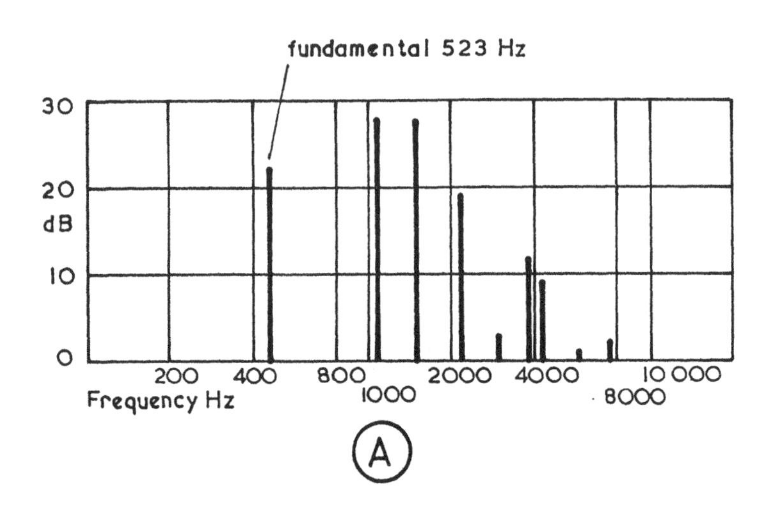

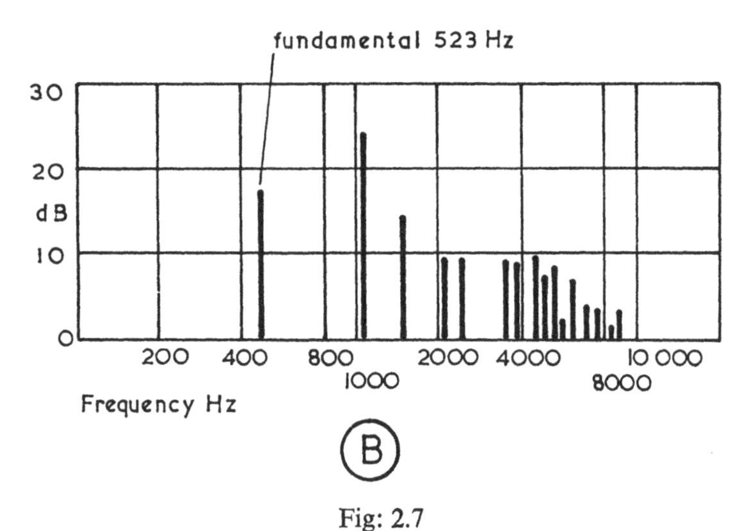

Fig: 2.7
(A) Simulated organ oboe voice harmonic structure.
(B) Real oboe harmonic structure.

produce the voices although in fact the electrical waveforms produced by the generators or their respective dividers may contain a large number of harmonics. On the other hand some tone generators, like the Hammond tone wheels, produce pure sine-waves which are completely devoid of harmonics. In this case harmonics of suitable pitch and intensity must be added to the fundamental in order to produce the required 'voice'. In the case of electronic tone generators and dividers, from which the waveforms may be complex i.e., containing many harmonics, the unwanted harmonics must be extracted and the wanted harmonics increased or reduced to the required levels. In this latter case the required harmonic structure cannot always be precise and it is for this reason that the voices may be nearly, but not quite exact. Exactness in voicing is costly and is only found in the more expensive organs. An example of a nearly but not quite exactness of voicing is shown in Fig: 2.7A which is the harmonic spectrum of the oboe voice of a typical electronic organ. The harmonic spectrum of a real oboe is shown in Fig: 2.7B. Further

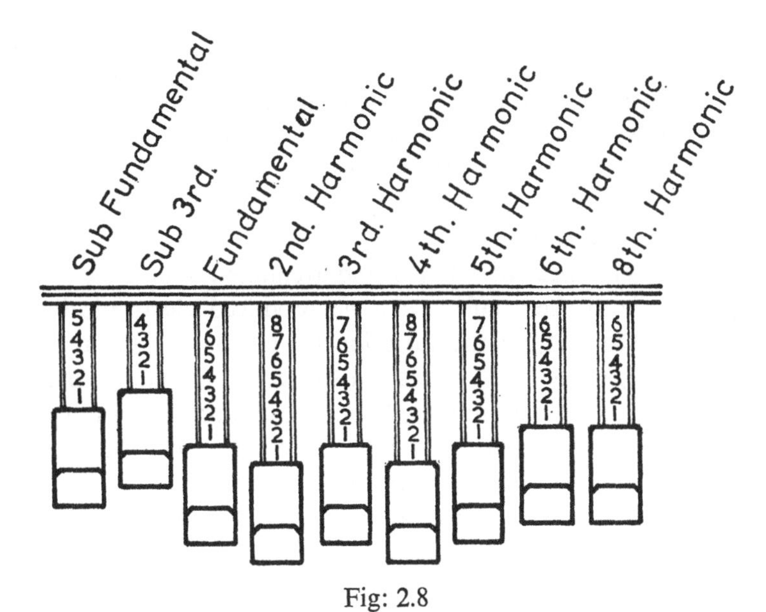

Fig: 2.8
The Hammond drawbar fundamental and harmonics selector system.
(*courtesy of Hammond Organs Ltd.*)

information concerned with the harmonics of musical instruments will be found in Chapter 1 which also deals with basic tone formation i.e., the elimination of unwanted harmonics from given waveforms in order to create a new one for the purpose of synthesis. The basic principles outlined are used in electronic organ voicing circuits although it should be noted that circuitry employed in any one make and type of organ may have been specifically designed for that instrument and may not necessarily apply to others.

Voicing by direct harmonic selection is common to certain types of organ and notably to Hammond organs which have the tone wheel generator system (described in Fig: 2.3) and their 'harmonic drawbar' method of selecting separate harmonics. The tone wheels generate almost pure sine-waves and from these, the drawbars select the fundamental, i.e. the basic pitch of the note being played and the requisite harmonics up to the 8th, but not including the 7th. The numerical setting of the drawbars determines the intensity of the harmonics selected (see Fig: 2.8). The drawbar method also allows the selection of the sub 3rd harmonic and sub fundamental. With this system it is possible to create not only 'orchestra' voices but also typical organ voices and voices to suit the player's own taste. In addition to the manually operated drawbars specific voices can be selected by touch tabs or by conventional but reversed colour keys at the left hand end of the manuals.

Vibrato and Tremulants

The musical aspect of vibrato and tremulant was dealt with in Chapter 1. Both these effects can be produced electronically and the rate or frequency of the variation for either is in the region of 5 to 10 times per second. There are also electro-mechanical methods of producing a tremulant notably by means of a rotating loudspeaker or a horn rotating around a loudspeaker, which most organists call a Leslie system. This arises from the fact that certain types of rotating horn or speaker systems currently in use are due to Don Leslie and are manufactured and sold bearing his name.

Vibrato

The vibrato or rapid variation in pitch effect is possible with quite a number of musical instruments and is an effect that is pleasing providing it is not over emphasised or over used. With acoustical musical instruments it has to be produced by the player and is part of the playing technique. Electronic organists have only to press a button to obtain instant and continuous vibrato and can instantly switch it off when not required. As most electronic organs employ oscillators for tone generation the introduction of vibrato or pitch variation is quite simple. All that is required is a slowly fluctuating voltage, i.e. a voltage which is varying sinusoidally at the requisite frequency of somewhere between 5 and 10 times per second. This can be generated with a single transistor employed as a phase shift oscillator as described in Chapter 1 (see Figs: 1.9 and 1.20). The sine-wave voltage generated by the vibrato oscillator is simultaneously applied to all the tone generators in the organ. The depth of the vibrato, which is determined by the amount of frequency shift, and the rate of vibrato, are both controllable and most organs feature controls for variation of both. Tremulant or loudness fluctuation is obtained electronically in much the same way in that an oscillator, usually a phase shift type, is employed to generate the tremulant control voltage. In this case, however, the voltage is used to control the level or amplitude of the signals from the tone generators or more likely the output level from one of the final organ amplifier stages. Tremulant produced by loudness variation is, however, not so musically pleasing as vibrato due to frequency or pitch variation, but various experiments with combinations of both have proved that the ear is very susceptible to either frequency or loudness variation of harmonics only. In this case no vibrato or tremulant is given to the fundamental.

Another rather unusual but nevertheless pleasing tremulant is that produced by acoustical/mechanical methods or perhaps more correctly electro-mechanical methods which employ either rotating loudspeakers or horns, or sound deflectors which rotate around loudspeakers. However, before dealing with these the tremolo or repeat effect must be mentioned. This is an offshoot of the electronics used for vibrato and tremulant and one which causes a note to be repeated rapidly much like the tremolo effect favoured by players of the mandolin. In this case the note signal, which would otherwise be continuous with the key held down, is rapidly cut on and off by an electronically generated recurring voltage, usually produced by a multi-vibrator as outlined in Fig: 1.21 (Chapter 1). On most organs the 'repeat' rate can be varied from very slow, about one repeat every half second, up to the speed used by a mandolinist for his tremolo, i.e. in the region of 10 times per second. The repeat rate control is greatly favoured by electronic organists as an aid to producing train sounds by running the repeat on close groups of notes. This is usually augmented with the sound of train whistles of various types derived from diminished chords and flute stops.

Electro-mechanical tremulants

The 'Leslie' organ tremulant systems are probably the best known but similar systems made by other manufacturers are given fanciful names like 'pulsation units' or wrongly called 'Doppler tremolo units' or are known by a trade name such as 'Spectratone'. All these systems are very similar and may employ either rotating horns or sound deflectors as in Fig: 2.9 *following page 32* which rotate around a fixed loudspeaker or, as in Fig: 2.10, loudspeakers which rotate. The rather strange thing is that very few of the manufacturers of such systems could explain to you exactly how the tremulant effect is produced. Most will say 'Ah yes, its due to the Doppler effect'.

The Doppler effect, named after Doppler who theoretically proved its cause, is an acoustic phenomena whereby the pitch of a sound moving toward and then away from the listener becomes changed. In systems in which the loudspeaker itself is rotated there is some Doppler effect, i.e., some apparent pitch change due to movement of the sound source but this is normally greatly masked by the rising and falling amplitude of the sound. Most of what is heard by the listener is in fact a loudness variation in the fundamental tones and their harmonics caused by phase differences between direct and reflected sound. This applies even more so to rotating horn and sound deflector systems which, because the sound source, i.e. the loudspeaker is fixed, do not produce any Doppler effect at all. Any apparent pitch variation from these systems is more likely due to large fluctuations in loudness as can be seen from the oscillograph shown in Fig: 2.11 *following page 32* which shows the rising and

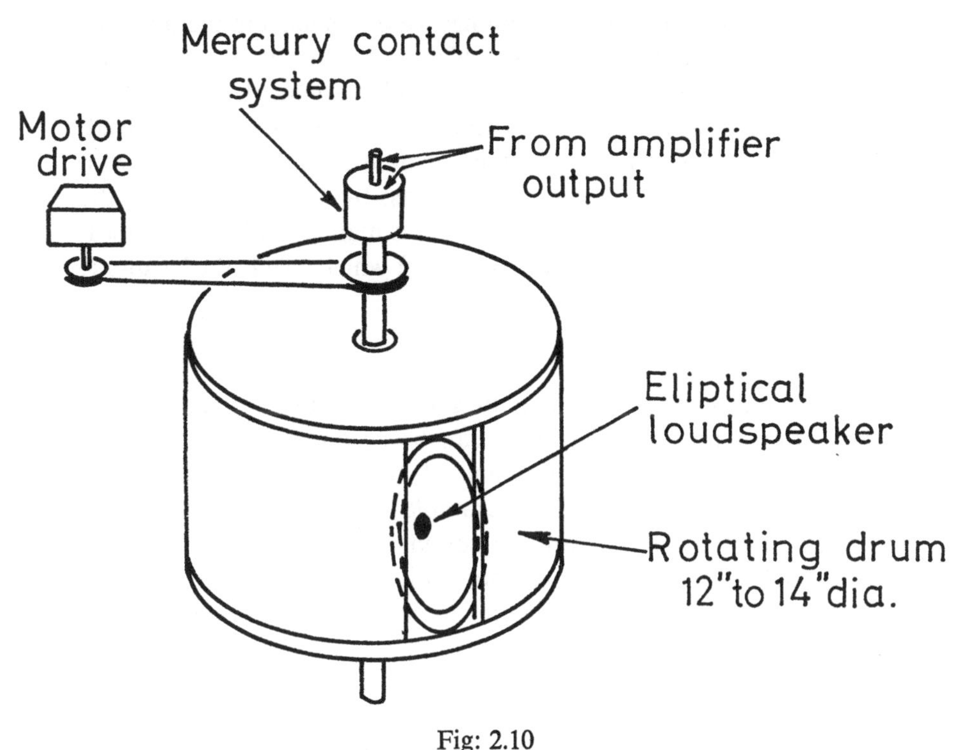

Fig: 2.10
Rotating speaker/drum tremulant system used in Lowery Organs.

falling amplitude produced by a rotating horn loudspeaker. The acoustic effect produced by rotating speaker systems is the same even when they are run at very slow speed to produce what is sometimes called the slow Leslie or Celeste effect. The majority of small domestic electronic organs are fitted with a rotating horn system like that shown in Fig: 2.9 *following page 32* which is from the Thomas Malibu organ. A rotating deflector system coupled with a rotating high frequency horn, shown in Fig: 2.12 *following page 32*, is used in one of the typical Leslie tone cabinets. One or two manufacturers employ a system consisting of two small loudspeakers which spin on a rotating arm, whilst yet another variation is a loudspeaker mounted on the periphery of a rotating drum (used in the Lowery Lincolnwood) (Fi:g 2.10).

Recently there have been many new developments in methods for obtaining vibrato and tremulant effects and one is a system that enables the player to obtain vibrato with varying degree by controlling the pressure applied to the key of the note being played. The Hammond method for obtaining tremulant from organs with tone wheel generators is also one in a class by itself and is called the Hammond Scanner system. The method is mainly electro-mechanical and involves the application of phase shift which effects the harmonic structure of the tone.

Special Effects

Electronic synthesis or imitation of the sound of almost any conventional musical instrument can be accomplished with a considerable degree of realism. Some of the more recent electronic piano instruments are a good example but with many electronic organs which have tabs for the sounds of other instruments, the 'imitation' is not always as realistic as it might be. Imitation of the sounds of known musical instruments involves generating the basic voice consisting of the fundamental pitch and harmonics, etc., and all other characteristics such as attack and decay and even pitch variation which may be due to the playing technique. Some instruments also produce random noise which is part of the whole sound. If any one of the tonal characteristics are missing then the voice loses its authenticity although it may still resemble the real thing. This is where imitation voices from electronic organs usually fail because not all the natural characteristics are produced and the result is something which may only vaguely resemble the real voice. However, even if the imitation is only a close approximation the listener will be satisfied. A sudden switch from the normal organ voices to even a poor imitation of a piano for example will convey the right impression as far as the average listener is concerned. This is a point worth remembering when buying an electronic organ that has tabs for the imitation of other musical instruments. A first hearing may convince you that they sound realistic, or nearly so, particularly when played as an immediate contrast to the normal voices.

Fig: 2.19 A typical modern plectrum guitar. The Gibson Les Paul custom guitar by Selmer Limited.

Fig: 2.20 Typical modern guitar amplifier and loudspeakers.

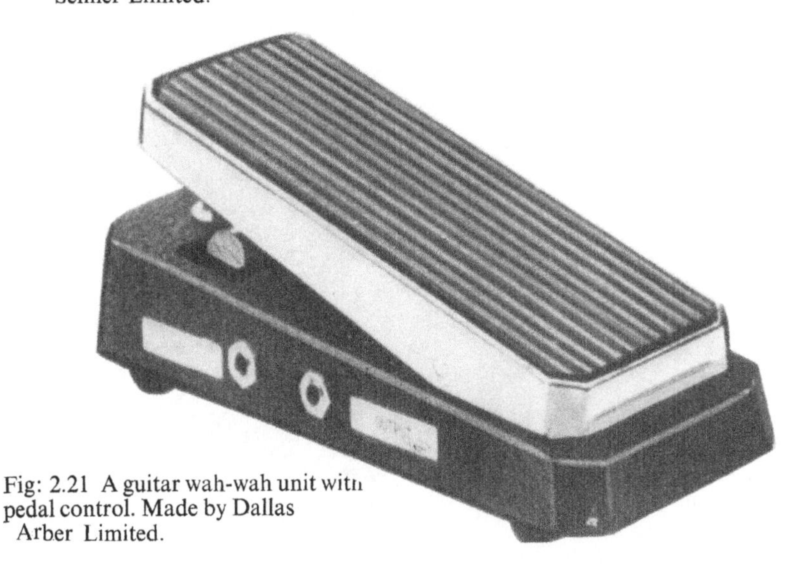

Fig: 2.21 A guitar wah-wah unit with pedal control. Made by Dallas Arber Limited.

Fig: 2.22 The Dubreq PianoMate. A conversion unit that turns a piano into an electronic organ.

Fig: 2.23 The Lowrey E.P.O. Electronic piano organ. Essentially an electronic piano that can be played as an organ.

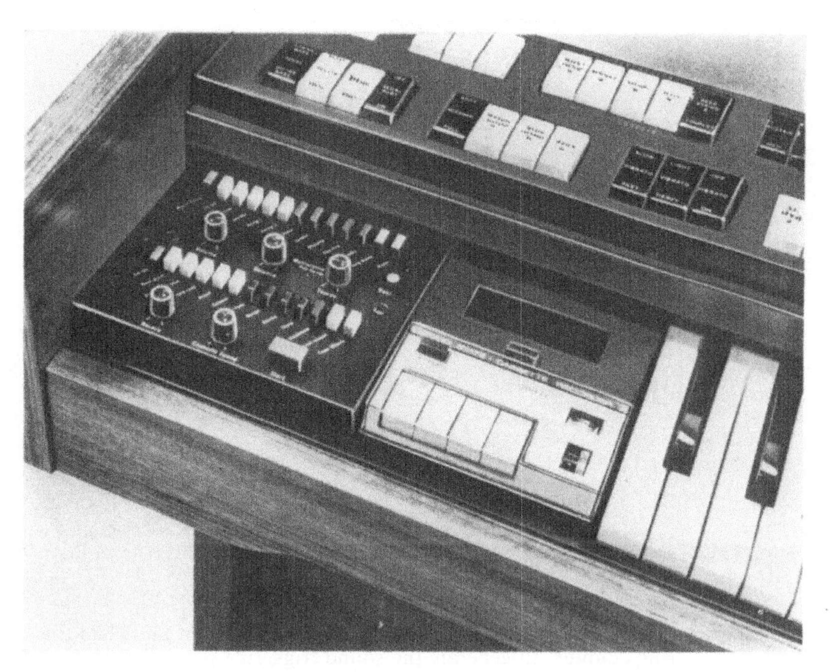

Fig: 3.1 The Lowrey K Factor. A built-in percussion-rhythm unit (left) and a built-in tape recorder (right) *(courtesy of Lowrey Organs Limited)*.

Fig: 3.2 Experimental percussion-rhythm unit built by the author (see text and Fig: 3.3).

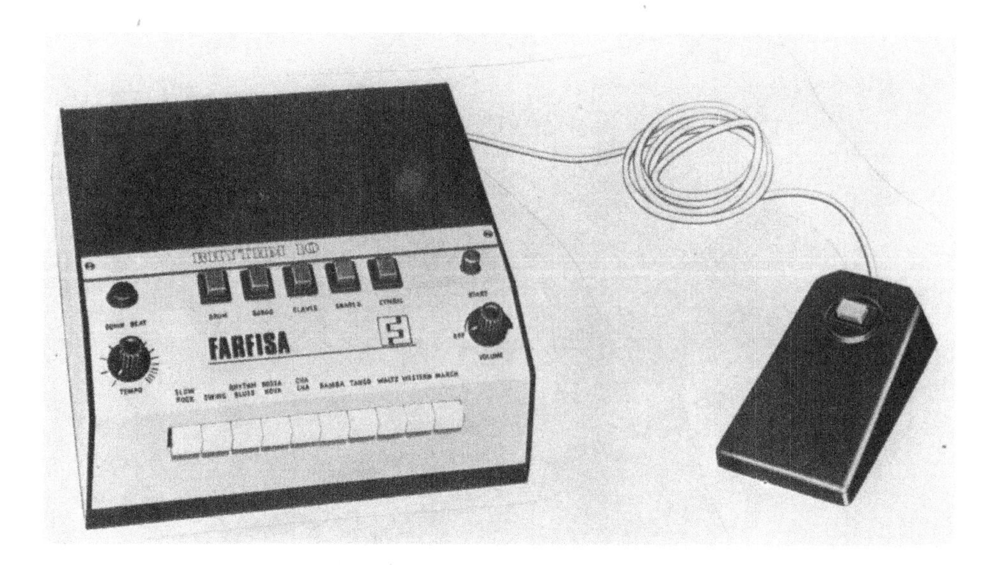

Fig: 3.4 The Farfisa Rhythm 10. Modern electronic percussion rhythm units like this employ ring counters to generate the sound triggering pulses.

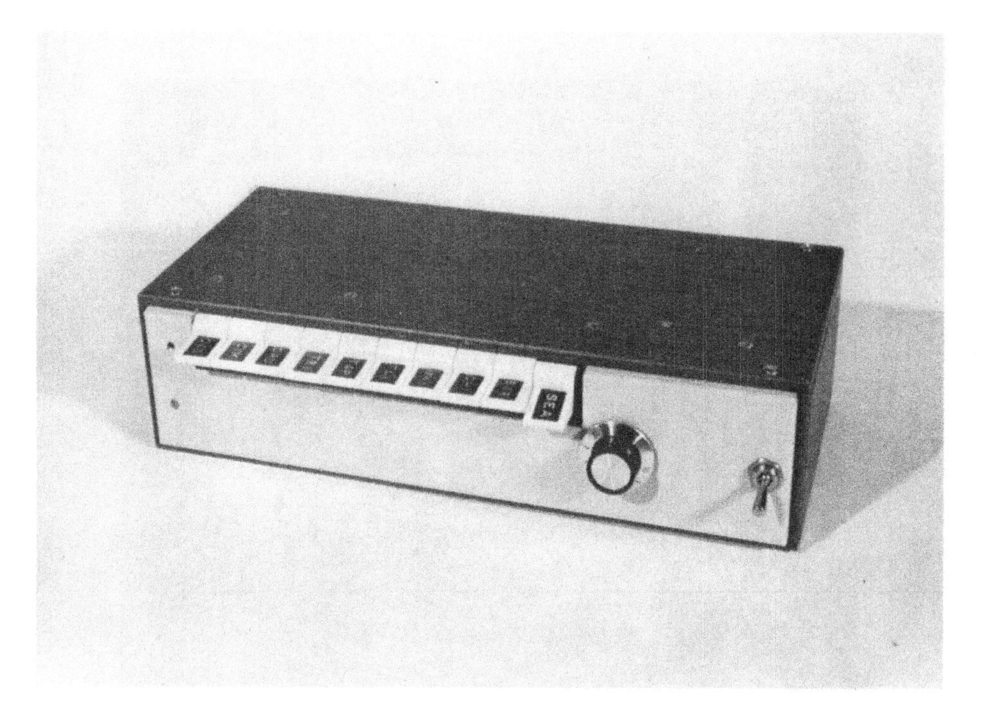

Fig: 3.14 Complete sound effects and percussion instrument generator constructed from the circuits in Figs: 3.5 to 3.11 *(photo by courtesy of Practical Wireless).*

Sustain and percussion

There are a great variety of circuits used for sustain and percussion which could be likened to attack and decay and for this reason are frequently used for imitation of other musical instruments. The percussive effect on normal organ voices causes the sound to end fairly abruptly whilst the sustain effect causes the sound to gradually die away of its own accord. On some organs the percussive effect is produced only when the key is depressed and held down, whilst on others, depending on the circuits used, the percussion may really be a short sustain. These effects are, however, often used in conjunction with certain 'organ' voices to produce rather unrealistic imitations of other musical instruments and it is not unusual to find tabs labelled 'piano' or 'guitar' for instance, which produce a sound composed of nothing more than one of the organ voices fed through a percussion or sustain circuit. Real imitation of musical instrument voices calls for pretty complex circuitry and is only likely to be incorporated in the more expensive range of organs. However, on many of the lower priced instruments, may be found some useful 'effects' ranging from pitch glide, which will produce something resembling the Hawaiian guitar glissando, to the 'wah' effect produced by trumpet players. These are all variants of either pitch shift, in which the control circuit alters the initial frequency of all the tone generators, or electronic control of the harmonic content of the voicing. Some organs even feature a 'bowing' effect which adds to the realism of an imitation 'cello or violin. Bell chimes are also a feature on many organs and these are usually obtained with the help of a sustain circuit and intercoupled groups of notes to provide the requisite fundamental and strong overtones. The 'glissando strip' to be found on some of the higher priced instruments is yet another electronic innovation that makes use of a sustain circuit and a thumb controlled keying strip usually fitted just below the upper manual.

Percussion instrument sounds

The most recent addition to electronic organs are built-in 'rhythm' units which provide reasonably lifelike imitations of percussion instruments such as drums, cymbals, claves and castanets, etc., and moreover 'play' these in a wide range of rhythms. Most, if not all of these systems will play automatically and have variable control over tempo. Some are not automatic but the generating circuits can be coupled to contacts on either the lower manual keys or the base pedals, or both, so that the percussion sounds are triggered by the player i.e., they follow the beat of the left hand accompaniment or the pedal bass. Some even allow the selection of individual sounds such as a crash cymbal for example, by the player, exactly when required and some are triggered into operation at the commencement of playing and stop when the player stops. All these units employ electronic tone generators coupled with attack and decay circuits and some include repetition voltage generators to provide drum rolls and the sound of castanets, etc.

A more recent development is the electronic generation of real sounds i.e., train whistles, taxi horns and ships' sirens, etc. Sound effects were a great feature of the cinema organs except that most of them were produced by mechanical means. Electronic sound effects generators of this nature are now being incorporated in a few organs and like the rhythm unit may eventually become a more or less standard fitting. The time may also come when electronic organs will more closely resemble the electronic music synthesizers dealt with in Chapter 4. Already there is a trend to combining small music synthesizers with electronic organs.

Teaching aids

These might be better described as 'learning aids' for most have been devised as an aid to learning to play the electronic organ without the help of a teacher. Many, however, provide little or no assistance at all and are used as 'selling aids', rather than learning aids. Advertisements and brochures will tell you that this or that patent self learning method attached to, or supplied with the organ, will enable you to play immediately. So in good faith you buy the instrument fully expecting to become a virtuoso overnight. Some of the methods, which include note charts to be placed on the keyboards, colour illuminated and annotated keys, simplified forms of written music, chord guide charts, etc., are no doubt useful aids to acquiring a knowledge of music and playing without the help of a teacher but they can never make you an instant competent performer. The ability to play, even reasonably well, stems only from enthusiasm and long hours of practice. There is, however, one aid to music learning and practice that quite a few manufacturers have begun to fit on electronic organs and this is the magnetic tape recorder. Equally a tape recorder not directly attached will do the same job and ways of using it to improve playing, etc., and which apply to all musical instruments are dealt with in Chapter 5.

Electronic organs—types and facilities

The few instruments shown in the illustrations are only a representation of the hundreds of different makes, types and sizes at present on the market. Prices range from a few pounds for a simple monophonic instrument like that shown in Fig: 2.13 *following page 32* to well over £5,000 for a theatre type organ like the Hammond X66 shown in Fig: 2.14 *following page 32*. The smaller domestic organs with two or three manuals and 13 bass note pedals come within the price bracket £500 to £1,000 whilst small single keyboard instruments and portables may be obtained at prices ranging from a little over a £100 to around £300 or £400. Many of the portables have no bass pedals and no amplifier system whilst others like the Philicorda shown in Fig: 2.15 *following page 32* have built in amplifiers and loudspeakers but no bass pedals. Some have a split keyboard so that the lower section (left hand) can be played with different voices or produce chords from single notes. The piano organ has a similar feature except that chord accompaniment is obtained from a series of buttons rather like a piano accordion. Some of the slightly more sophisticated domestic organs like the Conn Theatrette in Fig: 2.16 *following page 32* have two or three manuals and an extended range of voices and effects. Many include a percussion rhythm unit and even a small tape recorder as an aid to learning and practice.

Choosing an electronic organ can be very difficult because there is not only a great variety to choose from but also because each has its own particular range of voices and effects, etc. It is not simply a matter of buying an electronic organ just because it sounds like one. The newcomer may well find that any small cheap organ sounds good on first hearing particularly if he or she does not play. It is only after one reaches the stage of being able to play with reasonable competence that the shortcomings of a small instrument will be realised. It does not follow either that a small cheap instrument will suffice for learning, the object then being to get a larger and better instrument at a later date. If you decide to become even only a fair amateur player buy the best you can afford and choose carefully. Don't be influenced by sales talk or instant playing aids. Have lots of different instruments demonstrated if possible in company with someone who not only plays but knows a little about different types and makes. This should not be the shop demonstrator who may well be able to impress you with the sound and facilities of various instruments but who may not necessarily sell you one most suited to your pocket and requirements.

The Electric Guitar

The electric guitar has evolved into three major types, the plectrum guitar with six strings normally used for solo and chord playing, the bass guitar with four strings pitched essentially for the bass parts of music and the 'steel' guitar which may be called a Hawaiian guitar or pedal steel guitar according to its type. In both latter cases the word 'steel' indicates that the notes are selected by means of a short steel bar, usually round, which is used as a movable fret, i.e. the steel is employed to artificially alter the length of the strings. A pedal steel guitar has a foot pedal controlled tuning system that allows the player to change the otherwise fixed tuning of selected strings so that specific chords may be played without complicated manipulation of the steel. In all cases electric guitars employ an electro-magnetic pick-up that fits under the strings. As the strings are predominantly steel, any movement in close proximity to an electro-magnet will produce a small voltage across the coil. As shown in Fig: 2.17 the electro-magnet consists of a magnetized pole piece over which is wound a coil of fine wire. The voltage generated in the coil by movement of the string within the magnetic field of the pole piece, is an electrical representation of the vibration of the string, i.e. the waveform of the voltage will be according to the pitch of the string, the harmonics it produces and the attack and/or decay which may to some extent be controlled by the player.

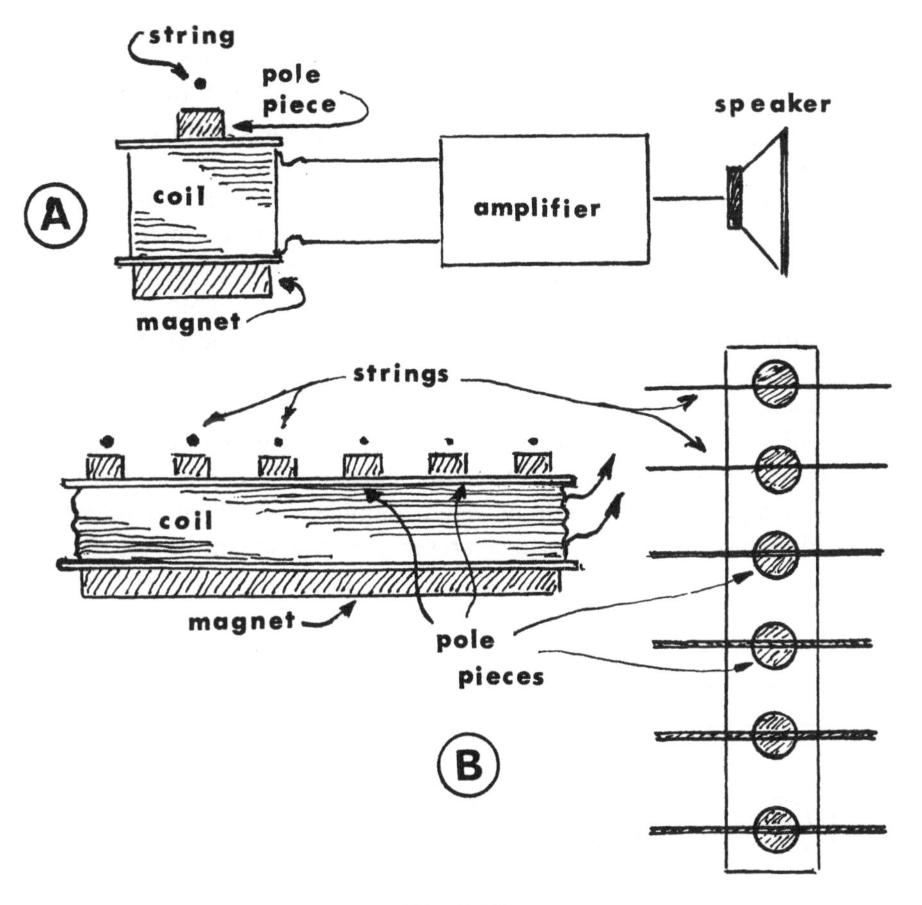

Fig: 2.17

Principle of the electric guitar magnetic pick-up. (A) Coil and magnetic pole piece convert string vibrations into electric currents which are then amplified. (B) Construction of a six string magnetic pick-up.

The plectrum guitar

Normally six strings are used although guitars are made with twelve strings which are arranged in pairs to produce the equivalent of six string tuning i.e., each pair is tuned in unison. The tuning, finger board and body of the modern electric guitar has evolved from the Spanish guitar which employs gut strings but which could be electrically amplified by means of a contact microphone. Acoustic plectrum guitars normally using steel strings can easily be converted for electrical reproduction by fitting a magnetic pick-up which is usually placed directly beneath the strings at the end of the finger board.

One of the special features of the plectrum guitar is the 'sostenuto' or sustaining of the vibration of the strings. Various methods are employed to maintain a good sostenuto the most common being longer strings than normal and bridges that do not damp the string vibration and are therefore usually made of hard metal. Specially modified bridges like that shown in Fig: 2.18 are also employed to maintain accurate tuning when the higher positions of the finger-board are used. Most electric guitars, like that shown in Fig: 2.19, *following page 64*, have built-in tone and volume controls, but are invariably used with high powered amplifiers which may provide additional tone control, artificial reverberation and vibrato, etc. Foot

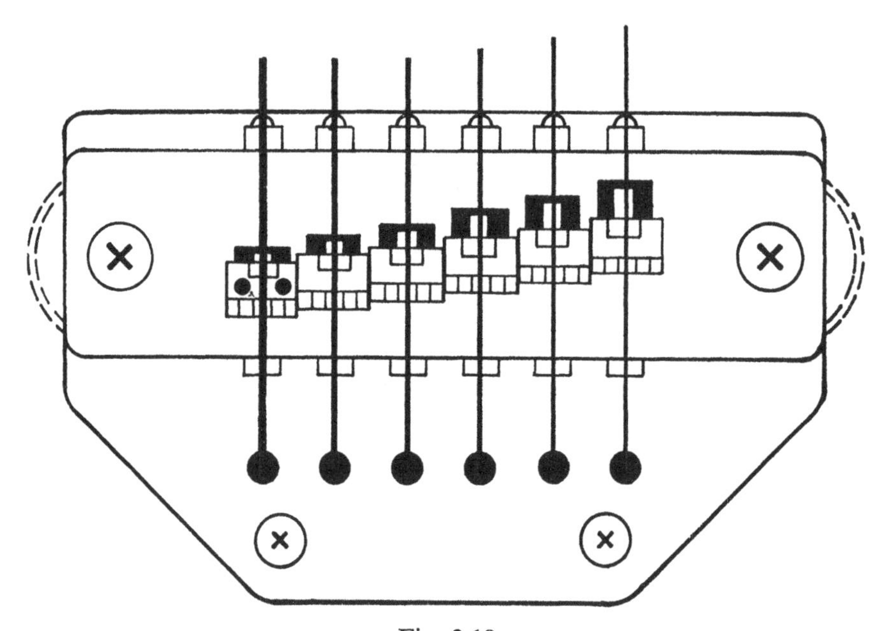

Fig: 2.18

Plectrum guitar adjustable bridge. Separate bridges, one for each string, allow individual tuning correction.

pedal volume controls are sometimes used to enable the player to alter volume whilst playing and to create a swell effect as the strings are played. All kinds of electronic devices have been produced to enable the guitarist to alter the tonal qualities of the instrument and to produce special effects known as 'fuzz', skying or phasing, wah-wah and unusual tremulant, etc. Fuzz is an effect produced by squaring the signals from the guitar pick-up and is achieved by first amplifying the signals and then using these to overdrive another amplifier. The result is a harsh sustained sound with a high degree of unrelated harmonics. Skying or phasing is usually produced by a special recording technique but can be accomplished electronically. The effect resembles the distortion created in radio signals due to fading and which consists mainly of slow variation in the harmonic content of the sound. Wah-wah is simply an effect resembling that produced by a trumpet player with a mute and amounts mainly to a very marked change in the level of harmonics i.e., the note sounded begins with the harmonic content greatly reduced and ends with full or exaggerated harmonic content. Incidentally most guitarists and indeed the makers of electric guitars and the amplifiers and various accessories that go with them, fondly imagine that the techniques and devices now used for tonal effects and tremulants, etc., are something new. Most of the effects already mentioned were possible more than forty years ago. The writer recalls designing a plectrum guitar amplifier which provided electronic tremulant, the wah-wah effect and so called fuzz, in about 1934. Experiments were also made at that time to halve normal guitar pitch by electronic frequency divider techniques.

The Pedal steel guitar

The 'electrics' of Hawaiian and pedal steel guitars are exactly the same as those employed for the plectrum guitar. The magnetic pick-ups are the same and all the effects that can be applied to plectrum guitar reproduction can also be applied to steel guitars. The pedal steel guitar is a derivation of the Hawaiian guitar and the method of playing it is similar. The pedal guitar has a number of foot pedal controls each of which can be used to mechanically alter the tuning of specific strings to enable the player to produce chords that would be difficult with fixed tuning. Most pedal guitars are designed so that the player is seated and some have twin necks, i.e. duplicated instruments on the same console, each with a different tuning and controlled by as many as ten foot pedals.

Guitar amplifiers

High power output is the main requirement and this necessitates large loudspeakers to handle it. Some amplifiers are built into the same cabinet as the speaker(s) and some are separate. Power outputs as high as 200 watts are not uncommon. Most of the smaller self contained amplifiers like that shown in Fig: 2.20, *following page 64*, incorporate volume and tone controls, a tremulant control (usually wrongly called a vibrato control) and have inputs for one or more guitars. Some also have provision for wah-wah and other effects such as reverberation. Many of the large amplifiers sometimes used by groups have a multiplicity of inputs for guitars and microphones, with separate tone and volume controls for different inputs and provision for reverberation and tremulant, etc.

Electric guitar accessories

This refers to accessories which are employed to modify the tonal qualities, i.e. to produce special effects such as wah-wah and fuzz, etc. The available range is vast and also includes reverberation and echo units that can be connected, like most other units, between the guitar and its amplifier. Wah-wah and fuzz units are mostly self-contained and foot operated like the wah-wah unit shown in Fig: 2.21, *following page 64*.

Reverberation units produce the 'cathedral' type reverberation in which the sound continues to reverberate for a fairly long period due to reflection from walls, etc. Practically all these units employ coiled spring delay lines along which the sound travels backwards and forwards, gradually decaying the whole time. Reverberation time can be adjusted from a small fraction of a second to as much as two seconds. The principle of operation is explained later in Chapter 3. Echo units are magnetic tape systems in which the signal is echoed by a tape head feedback path. The signals are recorded on a continuously running loop of magnetic tape and then picked up at a later time from the loop as it passes a separate replay head. The signals from this are then passed back (recorded) onto the loop only to be picked up again a fraction of a second later by the replay head. The direct and re-recorded signals are mixed into a common amplifier. The echo produced is an abrupt recurring type and quite distinctive from reverberation obtained with spring lines.

Hybrid electronic musical instruments

There are really few nowadays although the music synthesizer might be described as a hybrid electronic musical instrument. The Theremin made a brief appearance recently but this too is little used. The Theremin is an instrument that employed two oscillators each tuned to an almost identical frequency usually in the region of 100,000Hz. The frequency of one, however, could be altered by means of 'hand capacity'. This was done by having a vertical metal rod connected to the circuitry of one of the oscillators. As the hand of the performer was placed near the rod, the frequency of the oscillator changed. The difference in frequency between the controlled oscillator and the one tuned to a fixed frequency, determined the pitch of the note being sounded. Movement of the performer's hand thus controlled pitch. The output signal from a Theremin was usually fairly sinusoidal.

There have of course been various devices to make electric guitars produce organ-like sounds. The fuzz system in conjunction with a foot pedal volume control will do this. Making a piano sound like an organ is perhaps even more unusual although this could not be done electronically without the use of extremely complex circuitry and a magnetic pick-up for every string. A novel conversion device, however, was recently introduced by Dubreq Studios of London Limited and is called a 'PianoMate'. The keying section of the instrument covers a pitch range of four octaves and is attached to a piano as in Fig: 2.22, *following page 64*. The 'keying' contacts of the PianoMate' are actuated by the piano keys. The instrument is effectively an electronic organ capable of producing different organ like voices and consisting of the keying attachment and an amplifier system containing the tone generators, etc.

Electronic pianos that can be made to play as electronic organs are not uncommon. These were mentioned earlier but of those that were available until recently the Lowery EPO or electronic piano organ shown in Fig. 2.23, *following page 64*, is a good example. This normally plays as a piano with appropriate synthesis of either normal piano sound or harpsichord or honky tonk piano but can be switched to produce conventional organ sound plus pedal bass.

One other electrical musical instrument worthy of mention is the 'Electravibe' which is a vibraphone type of instrument in which the primary sound source is an aluminium alloy bar. There are thirty-seven of these which provide a three octave playing range and beneath each bar is a pick-up transducer to convert the normal acoustic tone to electrical current suitable for amplification. These instruments have electronic tremulant control and can be played via wah-wah and fuzz units in the same way as an electric guitar.

Last but not least is the electronic piano accordion. Normally a wind operated wind instrument, this too has been 'transistorized', even to the inclusion of its own power amplifier. Some, like the Farfisa Transicord de luxe, have provision for electronic percussion and vibrato, the wah-wah effect and reverberation and a whole range of voices, many similar to those obtainable with electronic organs. The Farfisa Transicord even features a rhythm generator with drum and brush cymbal. The power of the wind accordion is controlled to a great extent by movement of the bellows and this is part of the playing technique. So that this is not lost to those trained to play wind accordions, the bellows still operate with the familiar expand and retract motion but instead of pumping air, are used to actuate a photocell or LDR and light control system to increase or decrease loudness. The tone generators, dividers and voicing circuitry is similar to that used in electronic organs.

CHAPTER THREE

Synthetic Sound and Miscellaneous Circuits

Synthetic sound generators, which include percussion-rhythm units, are not normally classified as electronic musical instruments, although many generate single percussion instrument sounds such as those made by snare drums, cymbals and castanets, etc. The electronic percussion-rhythm generator not only produces the requisite sounds but 'plays' them in tempo and with a variety of rhythms such as slow 3/4 waltz, quick 3/4 waltz, quick-step, tango, etc., and various Latin American rhythms. These units generate snare and bass drum, castanets, clave, woodblock, brush cymbal and different bongo drum sounds which are normally played in rhythm. Some have buttons for individual sounds such as a crash cymbal to be produced as and when required by the user. Percussion rhythm units such as the Delsonic Auto-Drum Mk 2 and the Farfisa Rhythm 10 are self contained, except for a power amplifier and speaker, and are portable enough to be taken around and used with other musical instruments such as the piano and electronic organ. Many electronic organs now have built-in rhythm units like that shown in Fig: 3.1, *following page 64*, which can be started 'on the beat', when the organist starts to play or, like the Thomas Band Box, coupled to the organ lower manual and pedals so that the sounds are produced in tempo with the chord and bass accompaniment played by the organist.

Earlier percussion rhythm units employed a rotating drum on which contacts were placed according to the rhythm required. These contacts triggered off the requisite sounds and the drum was revolved at a speed proportional to the required tempo. An experimental percussion rhythm unit employing this technique is shown in Fig:

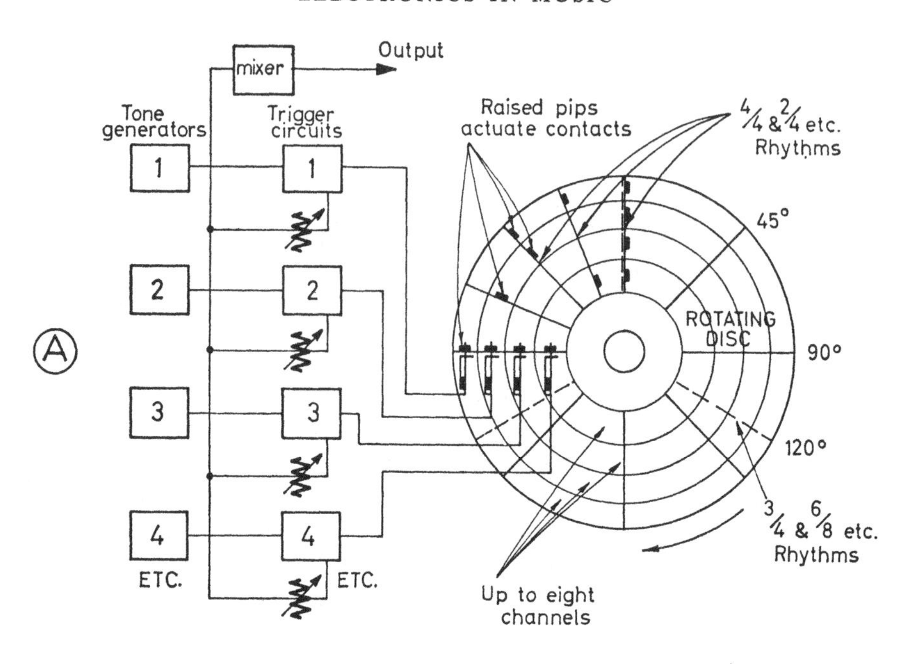

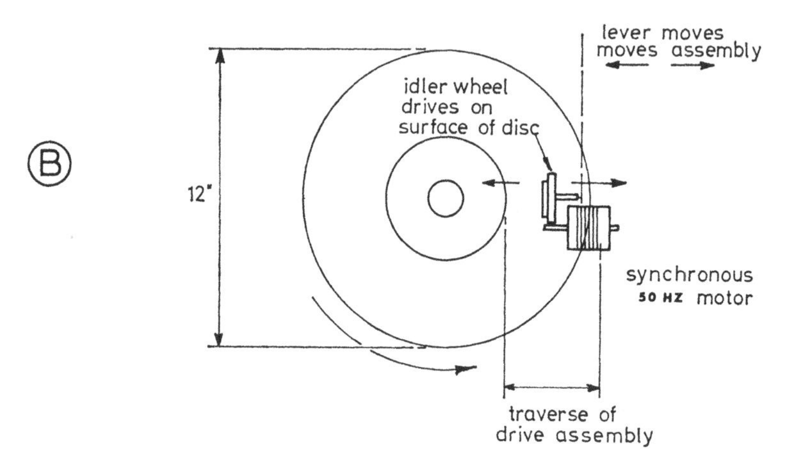

Fig: 3.3

Schematic of rotating disc system used in experimental percussion-rhythm unit shown in Fig: 3.2

3.2, *following page 64*. Small raised blocks set out around the rotating disc operated the contacts for triggering off the sound generators. The disc was turned by a synchronous constant speed motor via an idler wheel between the disc and the motor shaft. The motor and idler could be moved backwards or forwards between the outer and inner part of the disc and the point at which they were set determined the speed of rotation of the disc and therefore the tempo at which the rhythm unit performed. A schematic of the system is shown in Fig: 3.3. The contacts used for triggering off the sounds were standard gold plated electronic organ note contacts and the sound generators consisted of the following: 40Hz pure tone generator for bass drum, tuned pulse generator for tap sounds, i.e. woodblock and clave, etc., oscillators with inductive tuning for bongo drums at different pitches and white noise generators for drum brush, snare drum and cymbal sounds. The requisite attack and decay characteristics were also derived electronically.

Modern electronic percussion-rhythm units like that shown in Fig: 3.4, *following page 64*, do not employ rotating contact discs. The generation of the rhythm patterns is done entirely electronically by means of a 'ring counter'. This consists of a number of electronic circuits called bi-stables which conduct one way or the other, once only, when a triggering pulse is applied. Several bi-stables may be used each one connected so as to trigger another in a given order. The inter-connections are such that the initial triggering pulse which sets the first beat is passed on around the chain of bi-stables, hence the name ring counter. Each bi-stable generates the pulse required to trigger off a sound, e.g. bass drum or snare drum, etc., according to requirements. The last bi-stable in the chain re-sets the first ready for the first beat of the next bar. The initial triggering pulse is produced by a free running multi-vibrator (sometimes called a clock counter) the frequency of which is variable and can thus be set for any required tempo. The interconnections of the bi-stables are arranged so that counts of 4/4, 2/4 or 3/4 and 6/8, etc., can be obtained and the percussion sound generators are routed so that they can be switched to produce various combinations of sounds on different rhythm patterns. For example a common 4/4 rhythm might be arranged as beat 1 bass drum, beat 2 drum brush, beat 3 bass drum and beat 4 drum brush. The permutations possible with this system allow for a wide range of rhythm patterns.

Sound effects

The giant cinema organ is now regarded somewhat as a novelty but most of them were equipped to produce not only the sounds of percussion instruments but sound effects as well. Percussion sounds were usually derived from mechanically operated snare drums, cymbals, castanets and temple blocks, etc. Sound effects such a strain whistles, motor car horns, bird whistles, bell chimes and gongs, etc., were produced by either mechanical or wind operated devices. Electronic percussion-rhythm units, as already described, have become an integral part of modern electronic organs but few will provide individual sound effects. Some electronic organ manufacturers do, however, include one or two 'sound effects' such as a train whistle or chinese gong as extras. The sounds are generated entirely electronically and can be produced as and when required by the player.

Circuits for Synthetic Sound

It is not possible to include the full circuitry necessary for a ring counter percussion-rhythm generator because it simply could not be reproduced even in half a dozen pages of this book. Circuits of this nature are published from time to time in popular electronics technical magazines and references will be found at the end of the book. Circuits for single sounds, however, are not too complicated and some are included here for the experimentor.

Snare drum synthesizer

When a snare drum is struck two basic sounds are produced (a) 'strike tone' of the drumstick on the drum head, and (b) the noise of the snare wires. The generator circuit given in Fig: 3.5 produces both sounds as a continuous repetition, i.e. a drum roll whilst the key is held on. The transistors TR1 and TR2 between them form a multi-vibrator with a repetition rate of about 20Hz which is set by the pre-set control PR1. The pulse waveform from TR1/TR2 triggers

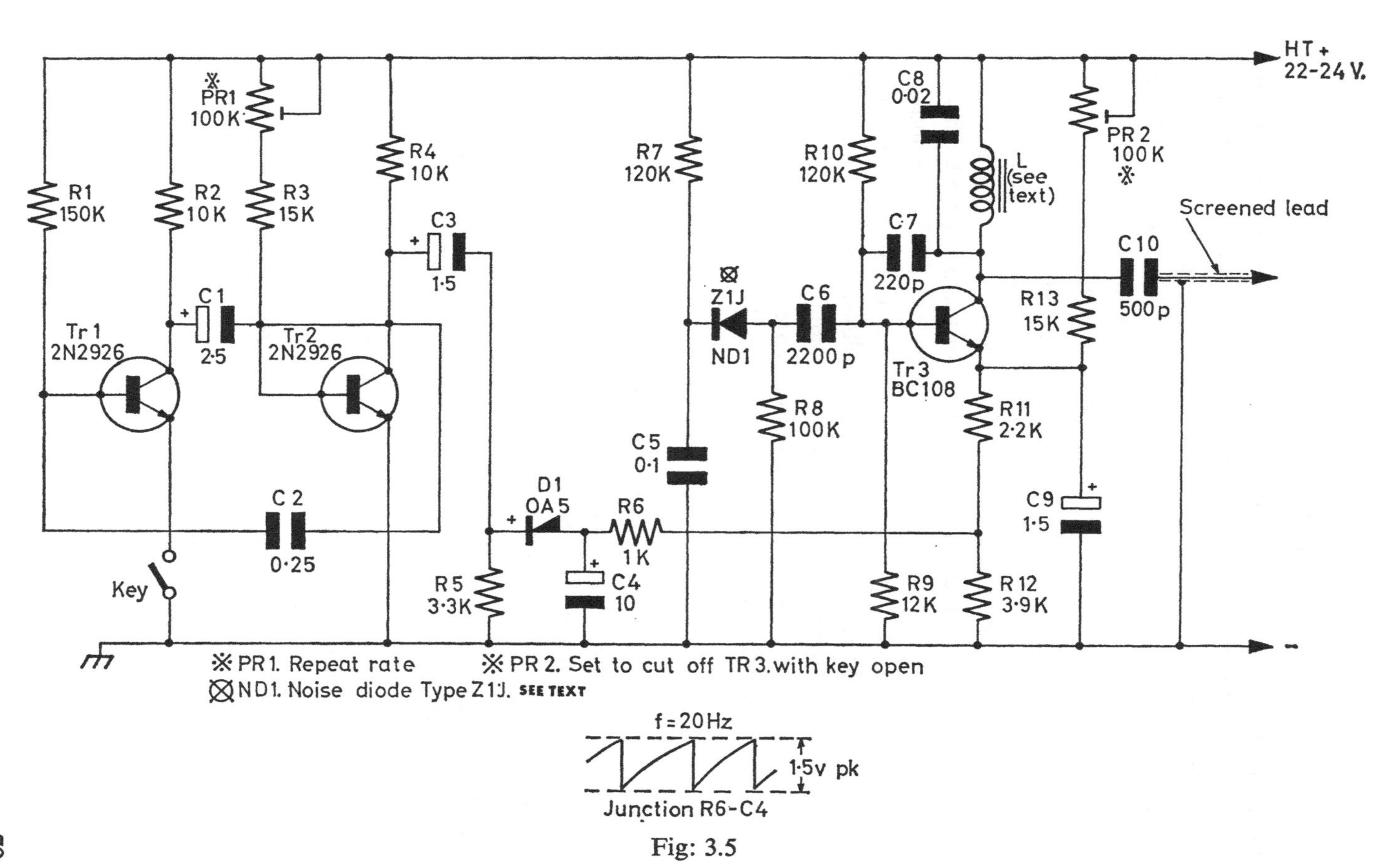

Fig: 3.5

Circuit for a snare-drum generator. Produces single or repeating drum rolls.

the amplifier/voicing circuit TR3. When TR3 conducts, noise from the generator (ND1) is amplified and at the same time filtered by the inductance L and capacitor C8. However, TR3 must be in a state of cut-off and this is achieved by the pre-set PR2. This should be adjusted until continuous noise from the amplifier output is just stopped. When the key is actuated TR3 conducts at the repetition rate of TR1/TR2. Normally this would only produce bursts of noise (the snare-wires). The strike tone is produced by ringing (part oscillation) due to the inductance which is further assisted by a small amount of feedback to sustain the ringing for a short period. This is provided by C7 between the collector and bass of TR3. The inductance L may be the primary of any small audio transformer but should have a dc resistance of about 600 ohms. The output signal level from the generator should be approximately 300mV.

Note: the noise diodes used in these circuits are type Z1J obtainable only from Semitron Limited, Cricklade, Wiltshire.

Circuit for synthetic cymbal crash

This generates a single cymbal crash when the key is pressed and consists of a triggering circuit which also provides the necessary attack and long decay. The circuit is shown in Fig: 3.6 and employs a Semitron Z1J noise diode as the primary tone source. The noise generator runs continuously and its output at C4 is connected to the amplifier/voice circuit TR2. This transistor is not connected to the positive supply rail and is therefore, not normally conducting. When the key is pressed a pulse is produced which immediately switches TR1 on so that C2 becomes charged to about two thirds of the supply potential. This allows TR2 to conduct at a decaying rate lasting for about two seconds. The noise passed and amplified by TR2 is filtered by the inductance L to achieve a slightly definable pitch. The circuit should require little or no adjustment but the decay time may be altered by changing the value of C2. The inductance L may be the primary of a small audio transformer and should have a DC resistance of 100 to 200 ohms. The output signal level should be in the region of 300mV at its peak, i.e. on the attack.

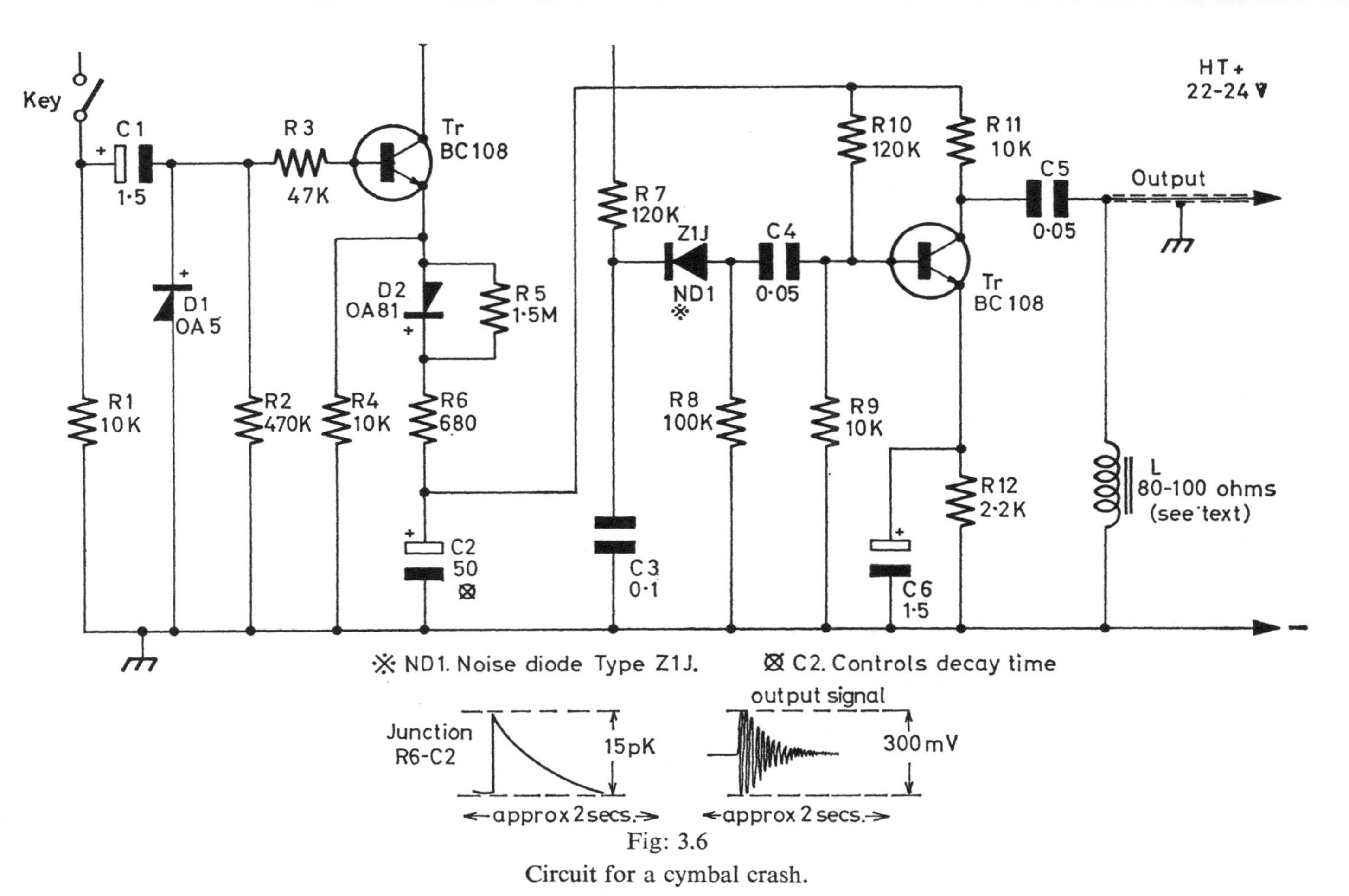

Fig: 3.6

Circuit for a cymbal crash.

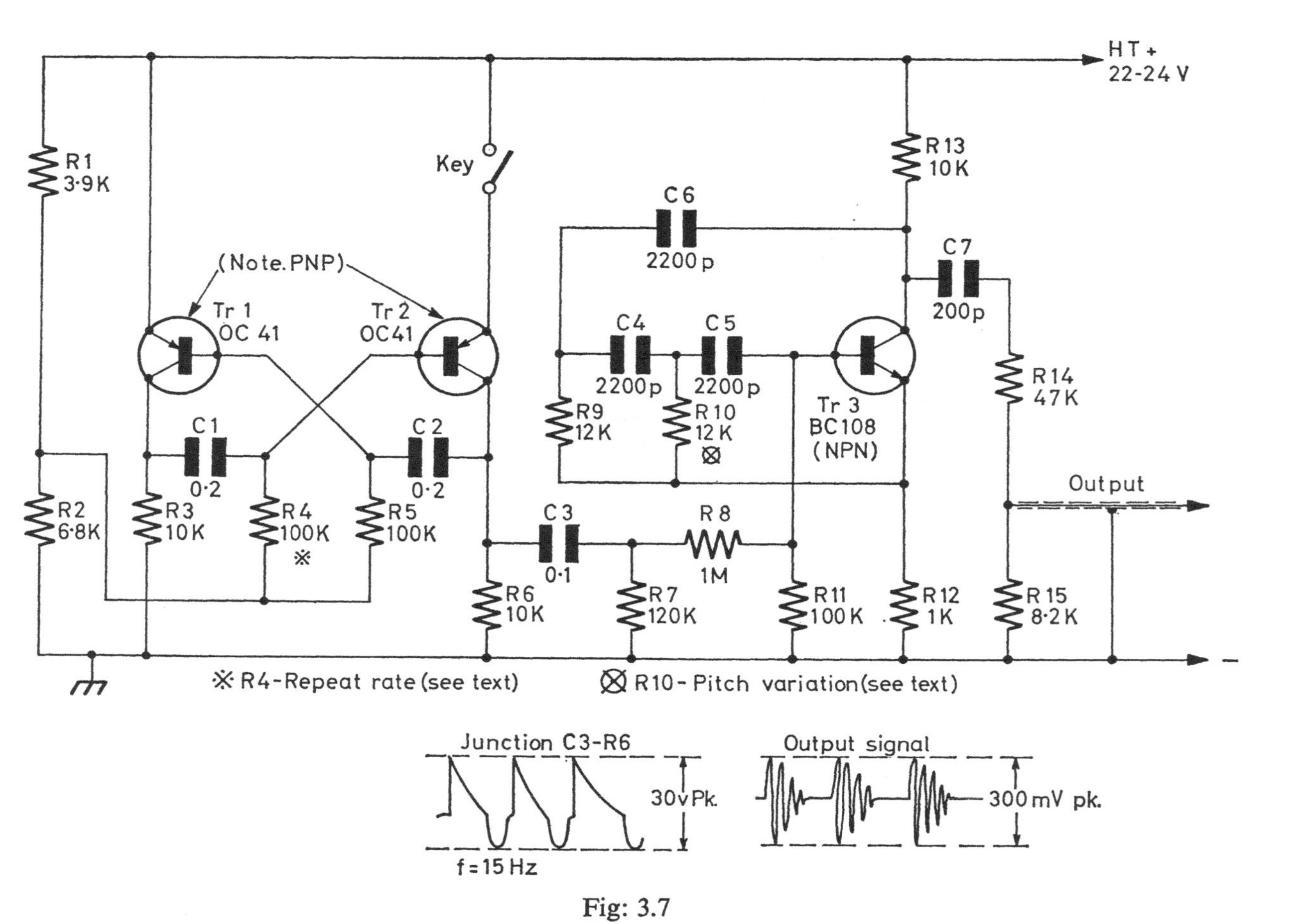

Fig: 3.7
Castanet generator. Produces intermittent or repeated castanet sounds.

Repeating castanets

This circuit as shown in Fig: 3.7 employs two transistors for the repetition control voltage and one for tone generation. TR1 and TR2 operate as a conventional multi-vibrator with a repetition frequency of about 15Hz but which can be adjusted by altering the value of R4 slightly. The waveform from TR1/TR2 is applied to the emitter of TR3 which is a phase shift oscillator biased to cut-off. This is tuned to oscillate at around 1500Hz and when triggered by TR1/TR2 produces the characteristic and slightly musical click of the castanets. The pitch of the castanet sound can be adjusted by slight variation of R10. The decay time is very short but can be adjusted by variation of R7. The output signal is approximately 300mV at the peak, i.e. on the attack.

Triangle Generator

The sound of that small percussion instrument, the triangle, is clear and high pitched and the waveform almost sinusoidal. The circuit in Fig: 3.8, therefore, employs a phase shift oscillator adjusted for a frequency of about 4500Hz. The output from the oscillator TR1 is first attenuated by R21 and then taken to the input of the control amplifier TR3. When the key is closed a pulse causes TR2 to conduct and generate a voltage at C9. The amplifier TR3 conducts only for the duration of the decaying voltage from C9 and thus provides the slow decay required for the sound. The sine-wave signal from TR1 does slightly overdrive TR3 and this helps to provide the characteristic metallic ring of the triangle. Note that a key click filter C7/C9 must be used in this case. The pitch can be adjusted by variation of R1 and the decay time by variation of C9.

Sound effects—Taxi Horn generator

This is a slightly more complex circuit and employs four transistors as shown in Fig: 3.9. The signal from TR1/TR2 is a typical multivibrator waveform which is modified and attenuated by the network R6, R7, C5 and R12. The pitch should be adjusted to between 200 and 250Hz by slight variation of R2 and the attack/decay characteristic by R15 and R16. Depression and instant release of the key should produce a short but typical ‘honk’ sound characteristic of

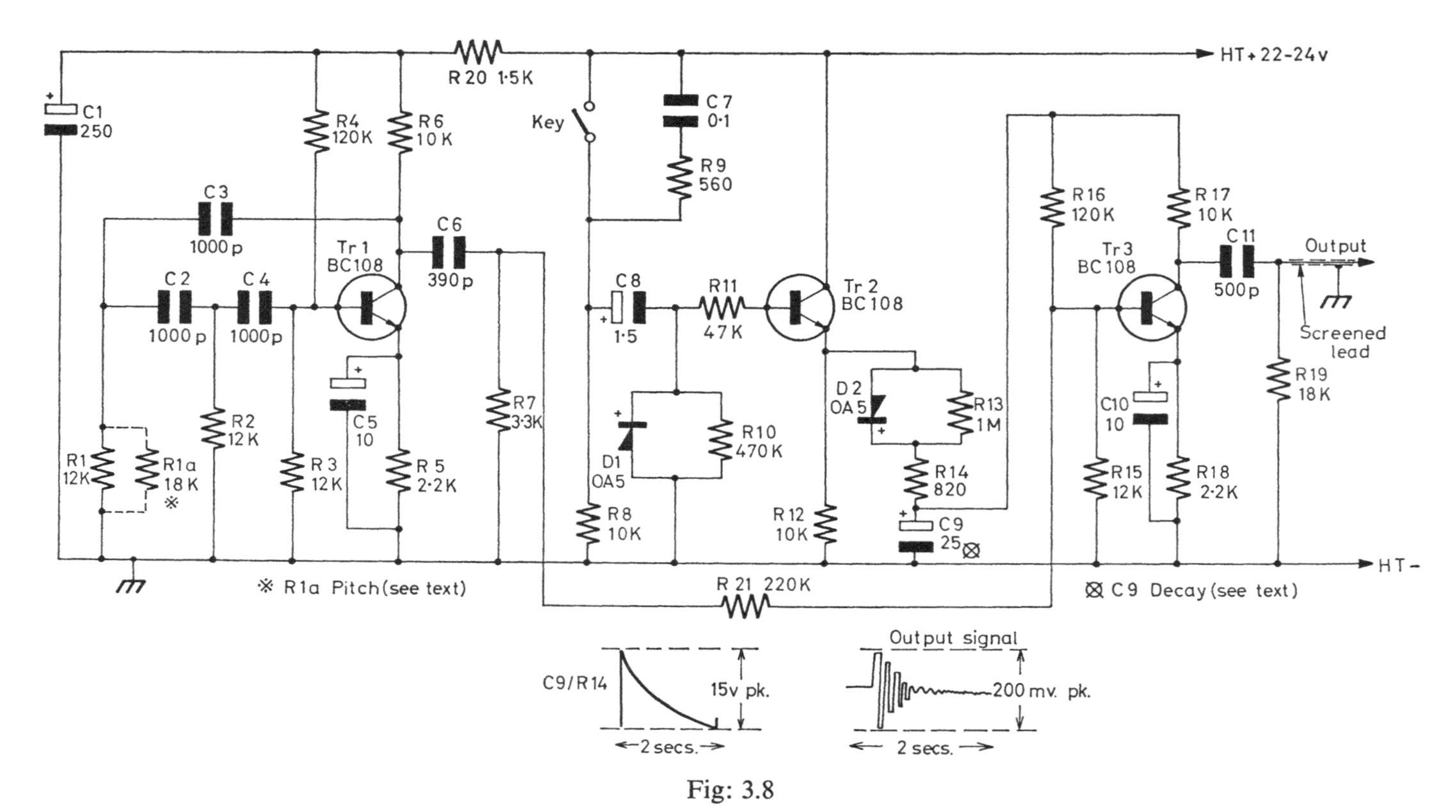

Fig: 3.8
Triangle generator.

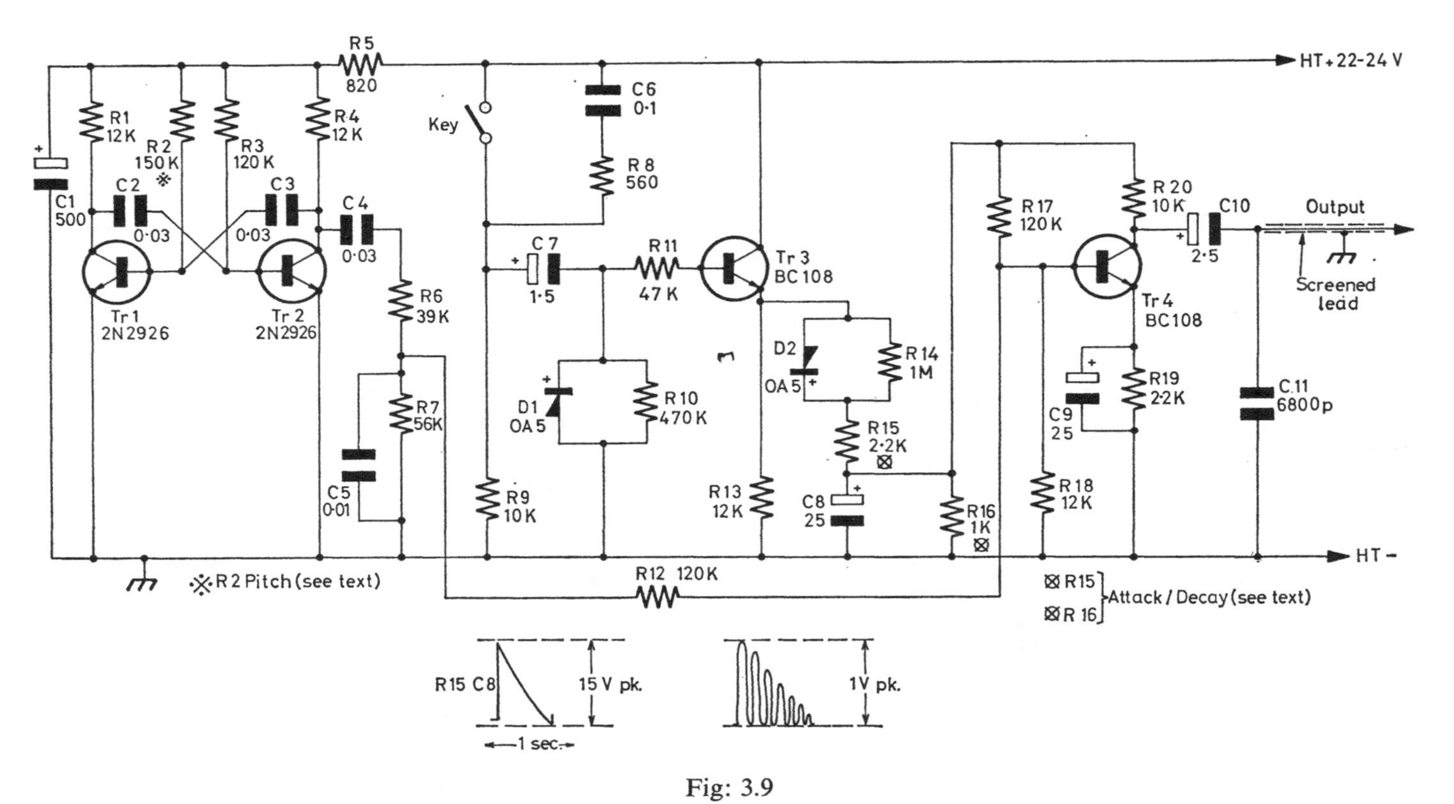

Fig: 3.9
Taxi horn (honk type) generator.

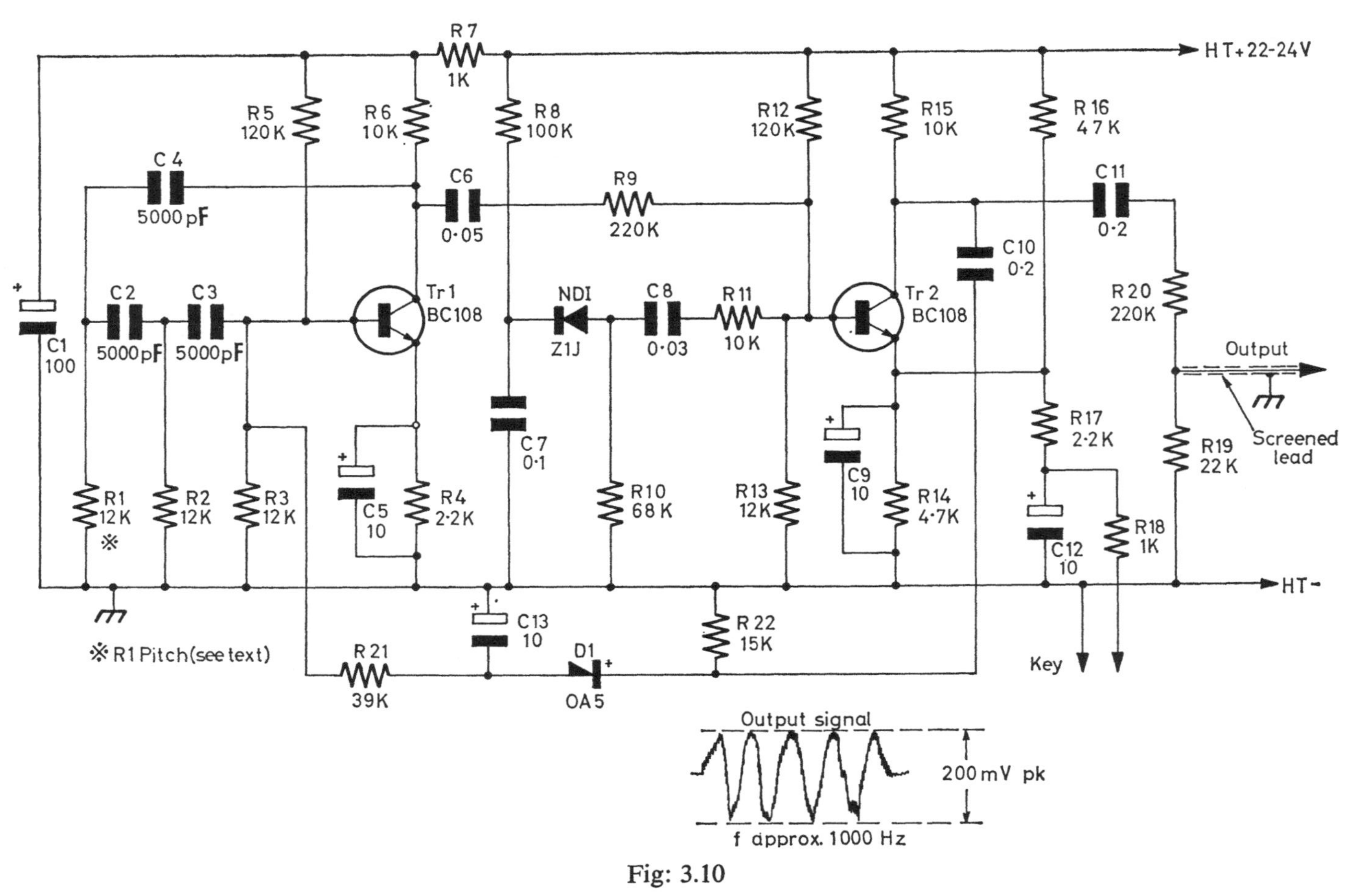

Fig: 3.10

Train whistle (steam type) generator.

old bulb type car horns. If the key is depressed and held down the decay will be very slightly longer but the sound will die away completely. The circuit is triggered by TR3 and the function of this transistor and its triggering voltages are the same as those (except for the decay time) for the triangle and cymbal circuits.

Train Whistle generator

Two transistors and a noise diode type Z1J are required for this circuit as shown in Fig: 3.10 and which produces a typical steam train whistle complete with noise content and pitch variation. The transistor TR1 is a phase shift oscillator the output of which is

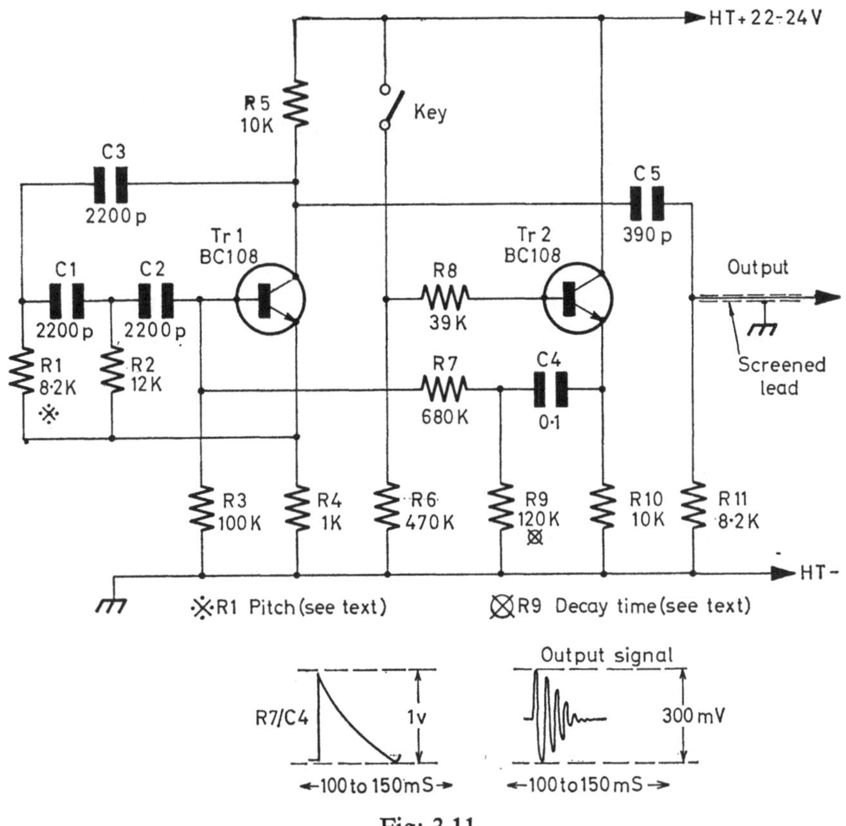

Fig: 3.11
Woodblock or clave generator.

attenuated via R9. The pitch should be approximately 1000Hz and adjustment to attain this can be made by slight variation of R1. The noise generator ND1 is the same as used for the cymbal and snare drum circuits and its output is mixed with the signal from the phase shift oscillator TR1. The control amplifier TR2 is normally biased to cut-off by R15 but when the key is closed TR2 will conduct and continue to conduct so long as the key is held closed. Train whistles of any desired duration can be sounded. The slight pitch variation is obtained by voltage control via the diode D1. The noise from ND1 adds the 'steam' sound to the whistle.

Woodblock or clave generator

This is a little less complicated and employs only two transistors. The circuit is shown in Fig: 3.11 and in this TR1 is a phase shift oscillator biased to cut-off and is turned on only when TR2 conducts, i.e. when the key is closed. The pitch and decay time are both very important if a realistic sound is to be produced. For instance if the decay time is too long the sound will be too much like a bell and if too short will sound like a click. Some adjustment of R9 may be necessary to achieve the right amount of decay. Pitch can be altered by slight variation of R1 which is nominally 8·2K. Adjustment to both decay time and pitch rather depends on aural estimation of the sound.

Circuit for Sea (surf on the beach)

Surf rolling up the beach and receding is a fairly slow movement and the sound itself, which is mainly random noise, rises and falls in amplitude at the same rate as the movement of the water. Here we need a slow but sinusoidal rate of change in signal level. The noise diode ND1 provides the white noise but to be effective a slight change of pitch in the noise is needed as well as the large amplitude variation.

The noise signal is fed directly to TR2 which is rendered almost non-conducting because of the large value resistor (1·5M) between the base and ht+. The base is, however, directly coupled (via R7) to the collector of the slow running phase shift oscillator TR1. As the potential at TR1 collector rises and falls TR2 will become fully

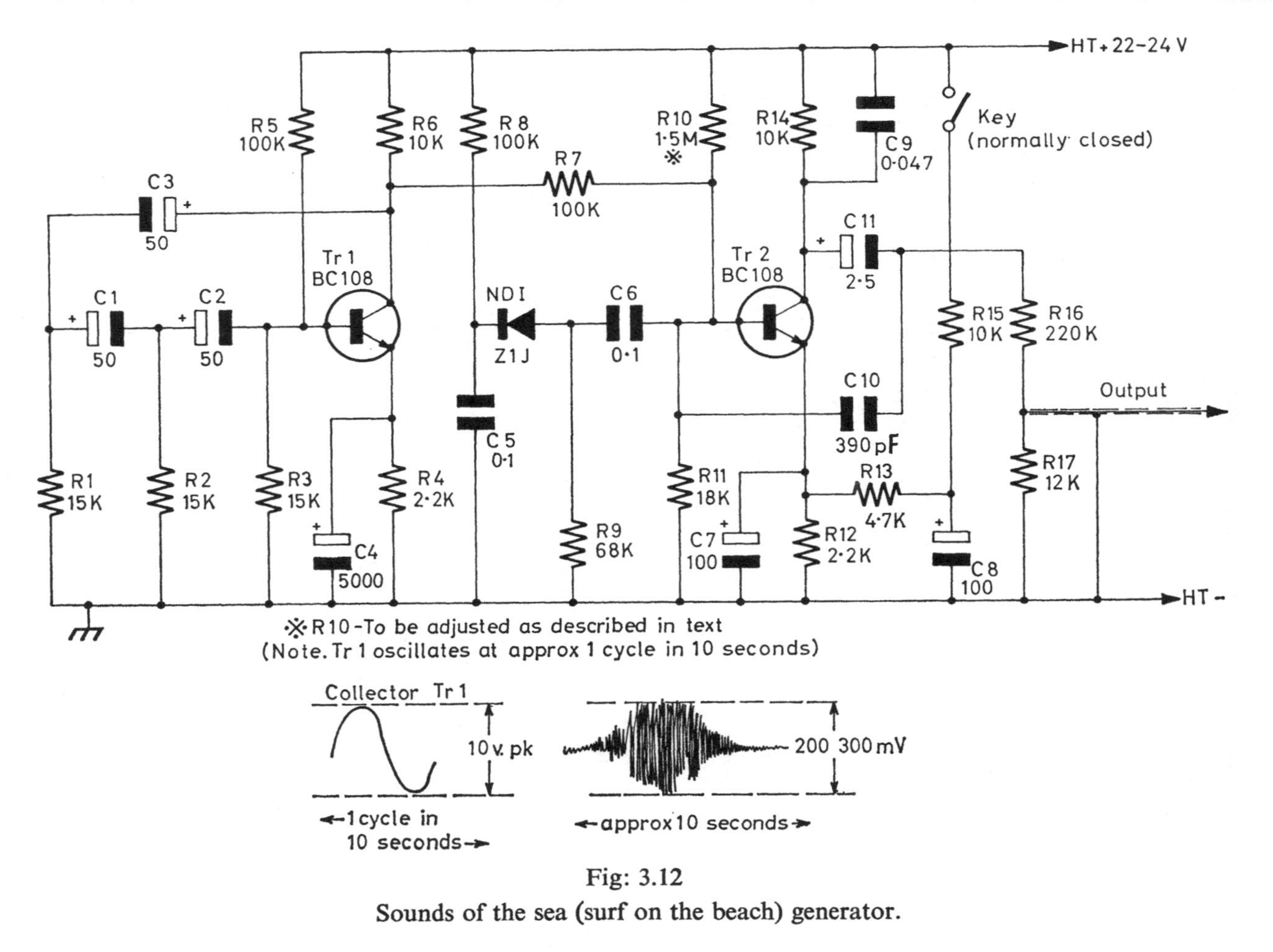

Fig: 3.12

Sounds of the sea (surf on the beach) generator.

conductive and/or quiescent at the same rate. The noise output signal from TR2 will rise and fall accordingly. The phase shift oscillator runs at approximately one complete cycle in ten seconds. The change in pitch of the noise output is produced by feedback via C10. The key is normally closed to hold TR2 in the cut-off state. When the key is opened the sound will continue until the key is closed again. In operation the sound must not completely disappear but simply rise and fall slowly. Adjustment to the value of R10 (1·5M) may, therefore, be necessary, i.e. the value may have to be reduced to 1·2M or increased to 1·8M. The sound produced by the circuit is quite realistic but should not be too loud when used as a background effect for 'desert island type music'.

All the circuits in the preceding paragraphs can be operated from a small power supply and their respective outputs connected to a suitable mixing circuit as given in Fig: 3.13. The original circuits were designed by the author and used to make a complete sound effects synthesizer as shown in Fig: 3.14, *following page* 64.

Electronic Metronome/rhythm generator

In addition to its realistic metronome sound, the generator circuit given in Fig: 3.15 will also produce a steady drum brush effect with variable attack and decay time so it also becomes a very simple rhythm unit. The tempo for either sound is continuously variable over the usual music range of about 40 to 250 beats per minute, i.e. from very slow (largo) to very fast (presto). The generator does not produce accented beats as this requires complex counting circuitry but as the beats are steady it is quite easy to play against them in three-four, two-four or four-four time, or multiples, depending on the setting of the tempo control. The drum brush attack and decay can be controlled so its 'voicing' can be set to produce a fast attack and short decay for fast tempi and soft attack and slow decay for slow tempi such as the slow waltz.

The generator can be operated from batteries or a small mains power supply. It takes only 2mA for the 18V ht rail so it would run economically from two PP6 (9V) batteries in series.

The first stage consists of a multi-vibrator, TR1 and TR2 which (please note) are pnp transistors with emitters up to the +18V rail.

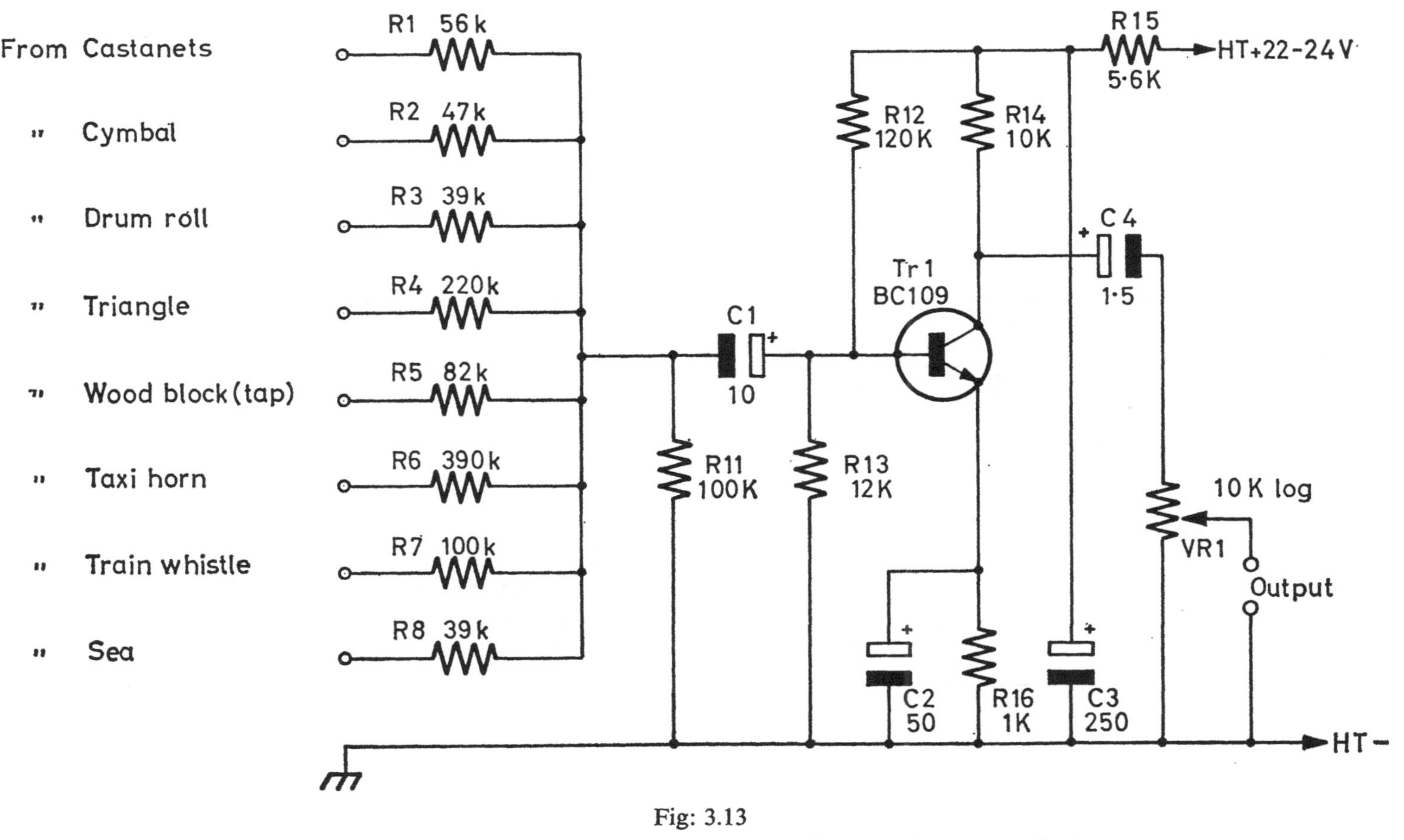

Fig: 3.13

Pre-amplifier/mixer for the sound generator circuits. Figs: 3.5 to 3.11.

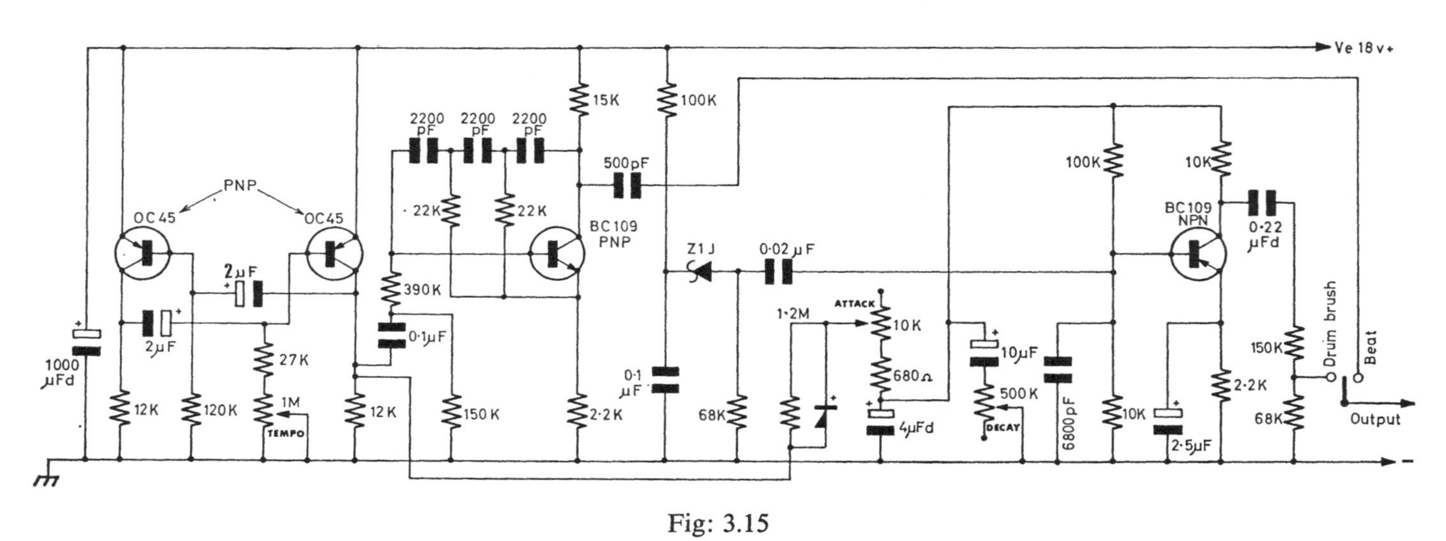

Fig: 3.15

Circuit for a metronome-rhythm generator.

The 'tempo' is controlled by VR1. The output from TR2 collector provides a positive going pulse which is coupled via C4 to the 'beat' voicer TR3. This is a npn type BC109 and is connected as a phase shift oscillator biased to cut-off. The positive pulse from C4 instantly drives TR3 on but the short time constant provided by C4/R5 allows the pulse voltage to decay quickly. The output from TR3 is, therefore, a short but gradually decaying burst of oscillation of about 1500Hz. The sound is a resonant tapping not unlike that of a clave and closely resembling the sound of a mechanical metronome.

The output from TR2 (multi-vibrator) is also fed to R15 and the diode D1 which, together with the attack/decay network VR2, R16 C11, C12 and VR3, provides a positive pulse with variable rise and decay time. This pulse from C11, drives the drum brush amplifier TR4, i.e. TR4 derives its ht potential from the pulse. The white noise for the drum brush voicing is obtained from the special Zenner diode Z1J. It must be emphasized that any 'noisy' Zenner may fail to produce the requisite amount of white noise which must be at least 50 to 100mV rms in level. (Suppliers of the Z1J noise diode are Semitron Limited, Cricklade, Wiltshire.) The outputs from the beat voicer (TR3) and the drum brush voicer (TR4) are taken to the changeover switch S1 which selects either voice. The signal output from this generator is approximately 1V rms and, therefore, more than sufficient to drive any conventional amplifier.

Echo and Reverberation

There is a subtle distinction between echo and reverberation although both are very closely related. In music reproduction each has its own particular application and meaning. Echo is an effect associated with sound reflected from high walls and buildings, etc., in the open air and consists of separate and distinct reproductions of the sound with a relatively short interval between each. Only a few echoes may occur, each one becoming progressively weaker in strength until they cease altogether. Reverberation is the effect usually associated with large empty halls and cathedrals, etc., and consists of multiple reflections of the sound with extremely short intervals between each, so short in fact that the ear cannot distinguish between them but instead hears what sounds like a prolonged single echo. The time taken for decay depends on the total strength and duration of all the reflections.

Echo and reverberation can both be produced with the aid of (a) magnetic tape recording, and (b) electro-mechanical devices. Both effects are used in conjunction with electronic musical instruments as well as in the recording of music and singing. Echo is favoured by electric guitarists and reverberation by electronic organists although either may use both effects. Few electronic organs are fitted with magnetic tape echo systems but many have built in reverberation units employing the most commonly used electro-mechanical device known as a 'spring line'.

Spring Line Reverberation

The speed of acoustical sound waves through metal is quite slow —around 5000 feet per second—slow that is by comparison with electrical sound waves through wires which is at the speed of light. An artificial reverberation unit may therefore consist of a direct electrical path with no delay and an acoustical path (of metal) which provides a delay time according to its length. As the individual echoes required must have a very short time interval between each the acoustical path does not have to be very long. However, the system must be able to maintain the repetition of echoes for a fairly long time—for up to at least two seconds. Metal plates and metal springs will do this and both are used for reverberation units. The sound must first be in electrical form and the electrical signals are then fed into a transducing unit in physical or magnetic contact with the plate or spring wire. Vibrations are set up along the plate or wire and continue to bounce back and forth until the energy is finally dissipated, i.e. the reverberation decays. The vibrations are, however, picked up at the far end of the plate or wire by a magnetic or similar pick-up device and reconstituted as electrical currents. These signals, which represent the reverberated sound, are mixed according to the strength required, i.e. the degree of reverberation, with the original signals. Metal plate reverberation units are of necessity very large and consequently are used mainly in recording studios. The spring line reverberation unit which consists of one or more steel springs around 12 to 18 inches long with appropriate driving and pick-up

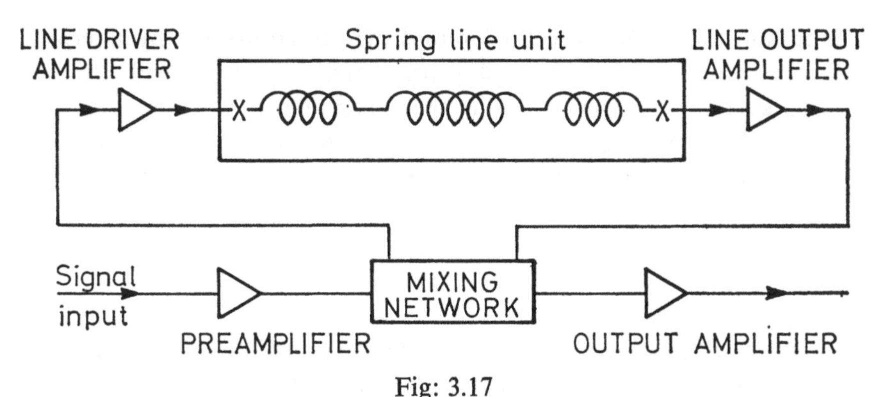

Fig: 3.17
Schematic diagram of a typical spring line reverberation unit.

units and pre-amplifiers is quite compact and can be built into an electronic organ or guitar amplifier or into a small self contained unit like the Grampian Model shown in Fig: 3.16, *following page 96*. This can be used with any amplifier system. A typical arrangement for spring line reverberation units is shown in schematic form in Fig: 3.17.

Tape Echo system

Magnetic tape echo units can be coupled to electronic and electrical musical instruments in much the same way as a spring line reverberation unit. Magnetic tape is the media for delaying the sound and the method used to obtain echoes is as follows. The sound, which must be in electrical waveform, is recorded directly onto magnetic tape. A replay tape head is placed a little further along the tape, i.e. beyond the recording head as shown in Fig: 3.18. This picks up the signals from the tape and from here they are returned to the recording head and re-recorded onto the tape. By careful balance of the returned signals separate and quite distinct echoes of the sound may be obtained which are then mixed with the original sound. The effect can be obtained with many domestic tape recorders, in fact in electronic musical instrument application, the tape echo unit is merely a simplified tape recorder. Most are quite small and self-contained like the 'Soundimension' unit shown in Fig: 3.19, *following page 96*. Most of these units employ a continuously running loop of magnetic tape and although the echo interval can be controlled by altering the distance between the replay head and recording head, this is often accomplished by changing the speed at which the tape passes the heads. Tape loops are, however, inclined to wear out quickly so some echo units employ magnetic discs. Multiple head systems are also employed to obtain an echo effect more closely resembling the reverberation obtained with spring line systems.

Electronic Tremulant

The principle electronic and mechanical methods for obtaining vibrato and tremulant for electronic musical instruments like the organ and guitar were outlined in Chapters 1 and 2. As already mentioned vibrato, as a pitch variation, is only possible electronically with instruments whose tone generators are electronic. With the electric guitar pitch vibrato can only be produced mechanically and

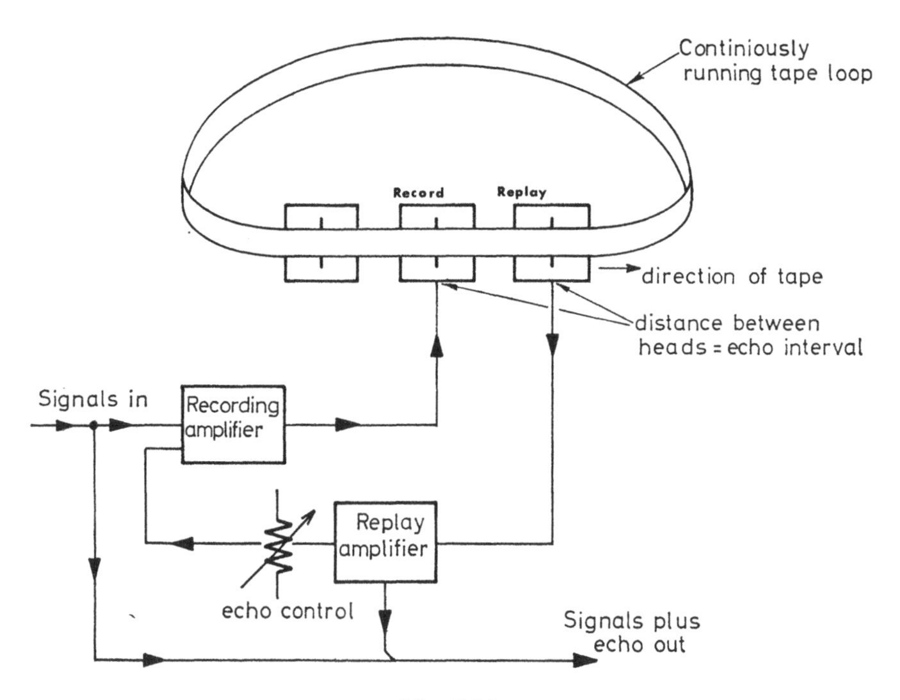

Fig: 3.18
The magnetic tape echo system.

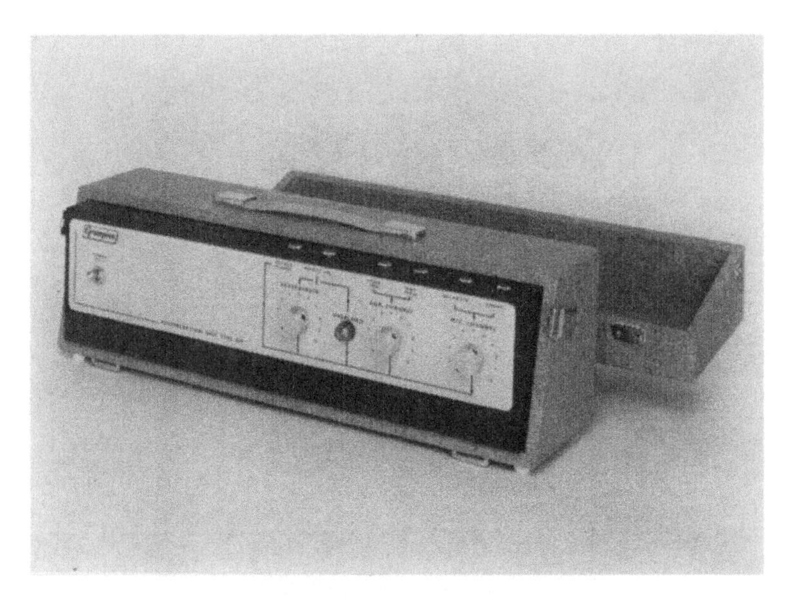

Fig: 3.16 The Grampian spring line reverberation unit.

Fig: 3.19 The Soundimension echo unit by Dallas Arbiter Ltd.

Fig: 4.6 Small section the BBC Radiophonic Workshop. A bank of keyboard controlled signal generators and a signal mixing unit.

Fig: 4.7 The Miller tone colour organ.

Fig: 4.8 Small music synthesizer (foreground) designed and built by the writer in about 1964.

Fig: 4.12 The Moog Mk3C Music Synthesizer *(photo by courtesy of Feldman Recording Limited).*

Fig: 4.14 The EMS Synthi. A portable mini synthesizer.

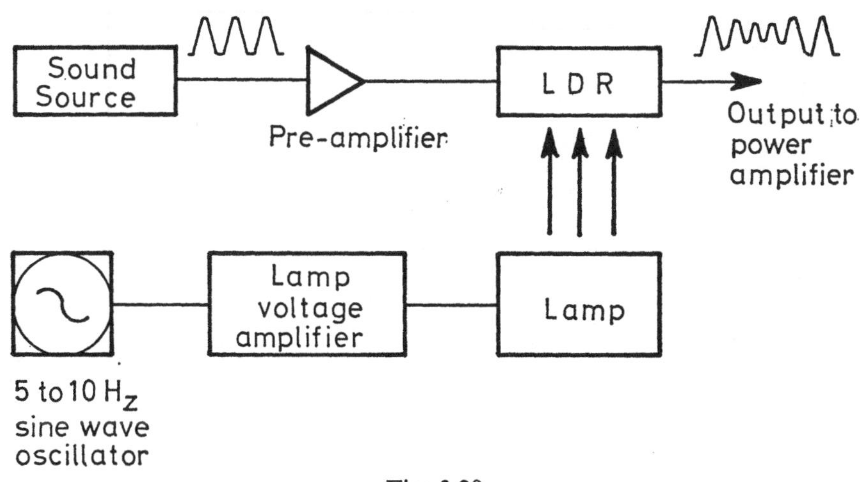

Fig: 3.20
Schematic of a tremulant system employing a light dependent resistor (LDR).

is normally accomplished by means of a semi-rotating bridge that alters the tension on all the strings. The bridge is rolled very slightly by means of a small lever which the player moves up and down at the required vibrato speed. The method is rather limited in its application for although the player can finger the correct notes, he cannot pick the strings and simultaneously move the vibrato control. Vibrato by this method must therefore be confined chiefly to sustained chords. The alternative is an electronic tremulant which can be applied continuously whilst playing. The usual method is to apply a low frequency sine-wave voltage to an amplifier so as to produce a varying gain and consequently a varying loudness. Earlier circuits for this were prone to producing thumping noises due to the harmonics from the tremulant sine-wave oscillator becoming amplified along with the guitar signals. One method of eliminating the thumping noise is to use a device that behaves as a volume control, i.e. one that is not used to directly control the gain of an amplifier. Such a device is the light dependent resistor (LDR) mentioned in Chapter 1 (Fig: 1.19) which can be used purely as a volume control, as will be shown later, or as a tremulant control in an arrangement like that given in Fig: 3.20.

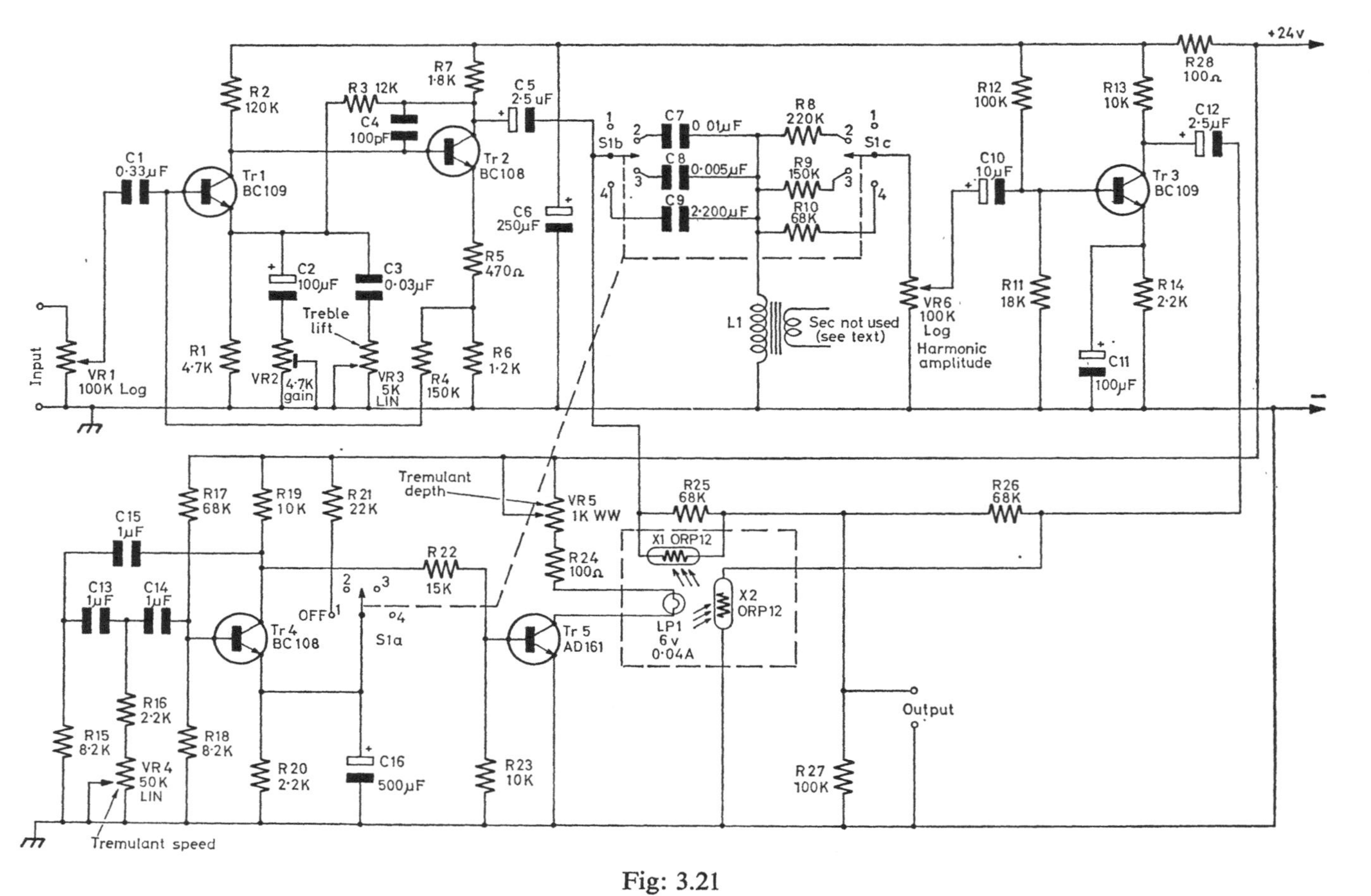

Fig: 3.21

Experimental tremulant generator suitable for electric guitars and organs using two LDR's and a filter circuit to produce a tremulant similar to that obtained with rotary tremulant loudspeaker systems.

Novel tremulant system

A fairly sophisticated tremulant circuit employing two light dependent resistors and which is capable of a tremulant similar to that obtained with rotating horn loudspeakers, is shown in Fig: 3.21. This is an experimental circuit and may require modification to produce specific variations in tonal control and/or to match up with other equipment. It could otherwise be used for electronic organs or guitars. The pre-amplifier, TR1 and TR2, has an input impedance of around 100Kohms and a sensitivity of approximately 50mV with VR2 at maximum resistance. With VR2 at minimum resistance the input sensitivity will be approximately 300mV. The output signal from the pre-amplifier is taken to a filter network and thence to the input of TR3 and also to the attenuator/mixing network comprised of R25 and the LDR marked X1. When the tremulant oscillator TR4 is operating the control lamp brilliance fluctuates sinusoidally and actuates X1. This produces a sinusoidal amplitude variation of the input signal. However, the input signal also goes through the filter circuits and TR3 which changes the harmonic structure in such a way as to emphasize some and suppress others, an effect produced by rotating tremulant speaker systems. The signals from TR3 are fed to the network comprising R26 and the LDR marked X2 which is also actuated when the tremulant oscillator is working. The signal outputs are combined across R27. When an electric guitar is played through the system the tremulant produced is much more interesting than the usual amplitude variation produced by conventional guitar amplifier tremulant systems. There is a continual variation of accented harmonics as well as fundamental tone. Note that all component values in the circuit are standard. The inductance L1 should have a dc resistance of about 600 ohms. The primary winding of a small audio transformer is suitable. The two LDR's must be mounted one each side of the lamp approximately one-eighth of an inch away. The lamp and the LDR's must also be completely shielded from external light or preferably mounted inside a light tight box. It should be emphasized that the use of an LDR for tremulant control completely eliminates the thumping effect usually obtained with direct control of an amplifier circuit.

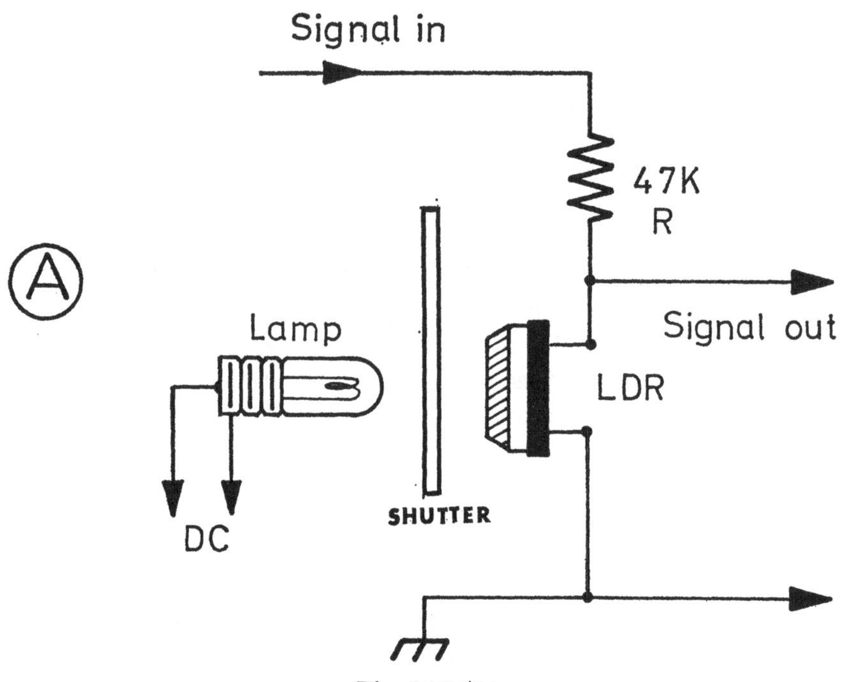

Fig: 3.22 (A)
Basic circuit for LDR volume control.

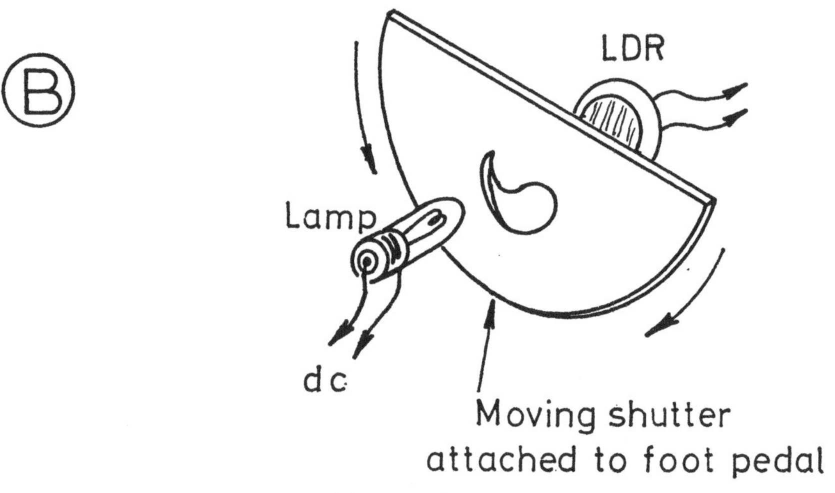

Fig: 3.22 (B)
Arrangement for control of light.

Light operated volume control

The light dependent resistor (LDR) can be employed as a volume control, is particularly suitable for foot pedal operation and has the advantage of being completely noise free. The main disadvantage of carbon track potentiometers, which are usually employed for this purpose is that they become noisy with continued use. The basic circuit for using an LDR is shown in Fig: 3.22A in which the resistor R and the LDR form a potential divider. With no light on the LDR its self-resistance is high, in the region of 50Kohms and as R is small the potential drop is small, so the input signal is practically unattenuated. When full light is on the LDR its self-resistance becomes very low, in the region of 100 ohms, which reduces the signal level to almost zero. The lamp must be operated from a dc potential to prevent hum modulation but may be an ordinary 6V flash lamp. The light reaching the LDR can be controlled by a moving plate (attached to a foot pedal) as shown in Fig: 3.22B. It is important that no external light is allowed to fall on the LDR.

CHAPTER FOUR

Electronic Music

In the opening paragraphs of Chapter 1 it was stated that 'electronic music' almost defies explanation because it is not the music which is electronic but the methods of creating it. In order to appreciate electronic music it is necessary to listen to it. Some like it, some don't.

The origin of electronic music goes back many years, in fact to the invention of the thermionic valve which could be made to oscillate at audio frequencies, i.e. generate tones in electrical form, and also to amplify them for reproduction via loudspeakers. In 1921 a concert of electronic music (?) was given in the Paris theatre de Champs-Elysées by an Italian Luigi Russolo with the aid of electrical sound generating and reproducing equipment. Really serious research in the field of electronic music did not really begin until after the 1939–45 war and one pioneer was Karlheinz Stockhausen, a composer with the WDR (West German Radio) studios in Cologne, who eventually set up a special studio for electronic music. Composers such as Henk Badings in the original Philips studio in Eindhoven, and Luciano Berio in the Milan Studio of Italian radio also contributed much to the development of electronic music. Studios came into operation all over the world and yet, until the invention of the music synthesizer, very little composed material was available on gramophone records although the output of the composers was quite large. One answer to this was given the writer by Philips recording studios in Holland who pointed out that the demand by the public for electronic music was just not sufficient to warrant the production of gramophone records. In view of some of the dubious *avant-garde* warblings produced by some composers at that time (*circa* 1960) this was not surprising. Although enthusiastic about electronic music, most composers produced nothing more interesting (for the public that is) than vast members of compositions based on 12 tone techniques, or which consisted of material without related pitch, scale or harmony. To untrained ears much of this music (?) sounded as a series of noises and tones of indefinite pitch or timbre without rhythm or harmony structure of any kind.

A few composers did, however, produce electronic music much more closely related to conventional music and which was based on the tempered musical scale, its related keys, pitch, harmony and rhythms, etc. They did what should have been obvious in the first place, and this was to apply the new tonal possibilities offered by electronics to the kind of music already understood and appreciated by the public at large. Here two composers come to mind who deserved more credit than they ever received. One is Tom Dissavelt, a Dutch composer who has produced many fine electronic music compositions with the rhythmic and harmonic structure of popular music but making the most of the unusual tonal characteristics that can be created with electronics. Another is Dick Raaymakers, formerly with the Philips studios in Eindhoven, who gained high awards for his work and who also retained the rhythm and harmony of music we understand. More recent contributions are those produced by composers and 'manipulator/players' using the full resources of the music synthesizer. Notable work is that by Walter Carlos which includes ten compositions by Bach performed on a Moog synthesizer.

Electronic Music Techniques

Until the invention of magnetic tape recording the production of electronic music presented many problems to the composer because the only means of storing (recording) electronically generated tones was the gramophone record. Some means of recording was essential as most electronic composition was created in parts, some often containing but a single note. The order of assembly also played a major part in composition, hence the importance of being able to record or store different parts whilst others were being produced. When all the parts were finished they were then assembled in the order required by the composer. When magnetic tape recording became available it provided a means not only for storage and easy reassembly, but also for further treatment of sounds. For example, a recorded sound could be reversed, so that its attack became the end of the sound instead of its beginning, by simply playing the tape recorded sound backwards. The tape recorder also allowed a series of sounds to be recorded and then replayed at a different speed thus changing the pitch of the sounds and the time interval between them. Moreover magnetic tape was easy to cut and splice which made reassembly of recorded sounds into a required order a comparatively simple task. This was something that was not possible with a gramophone record from which sound could only be put into a different order by re-recording but which also spoiled the quality quite considerably. However, even the advantages offered by magnetic tape permitted only fairly short compositions because the work of cutting and splicing together hundreds of separate pieces of tape containing the elements of long compositions was quite formidable. The writer recalls the production of an electronic music composition by Luciano Berio which lasted about three minutes when completed, but took six weeks to produce!

Nevertheless magnetic tape is still the prime media for recording electronic music, although a certain amount of tape manipulation is still employed, despite the facilities of music synthesizers and multi-track recording. Many of the earlier techniques can, however, be practised by students and amateur enthusiasts, using fairly simple equipment and with this in mind a good deal of this chapter is devoted to the sound sources, treatment and recording methods that can be used for study and experiment in electronic music and for the creation of special sound effects, etc.

Experimental Electronic Music

The tape recording enthusiast with a knowledge of music will find that the combination of magnetic tape and electronics offers some highly interesting and creative possibilities. It should be made clear at this point that electronic music does not preclude the use of key-board electronic musical instruments, in fact most of the professional studios use keyboard controlled electronic sound sources of one kind or another from giant music synthesizers down to simple monophonic (single note) electronic organs. Let us first consider setting up for the production of electronic sounds and musical effects rather than for composed music. The kind of equipment suggested is a tape recorder, one with a track-to-track recording facility and built-in-mixing is ideal, or better still two tape recorders with provision for mono or stereo recordings. A separate signal mixing unit is advantageous and then we require an electronic signal source with variable frequency control covering most of the audio frequency range. An audio signal generator with sine and square-wave output is preferable. A ring modulator is not difficult to make and is a device used in modern synthesizers to produce chords of tones from others fed into it. Various simple control circuits could also be built for producing special effects such as vibrato and controlled attack and decay. A noise generator is another signal source used for electronic music and is quite easy to make but before dealing with equipment in detail (circuits for various items will be given later) one may well ask, what exactly can be created or composed that is either electronic sound effects or real electronic music. The dividing line is very thin for one can quickly reach a point where it becomes difficult to differentiate between musical sounds or effects and composition which could quite legitimately be classified as music. Those trained in, or having some knowledge of music composition would no doubt be able to compose for electronic music. Almost anyone with a little knowledge of music and some degree of imagination could certainly produce material with a musical content that would be useful as a background to plays and films, etc. Lower in the scale are 'sound effects', the kind of thing used for science fiction or horror films. There is scope for everyone, however, including amateurs who can enter their efforts in contests like the annual British Tape Recording Contest which has a special category for experimental recording and includes electronic music and sound montage, etc. (See appendix).

Recording techniques

Tape recording techniques, as applied to the creation of electronic sound effects and music are best tabulated so that one may see at a glance what can be applied according to the equipment available. These are as follows, but are within the capabilities of tape recorders likely to be used by amateurs:

1. Normal recording and/or playback at a given tape speed.

2. Replay at speeds faster or slower than the recording speed.
3. Replay in reverse. Possible with full track and most stereo tape recorders.
4. Recording with tape head feedback (echo).

To these basic recording techniques can be added:

5. Dubbing from track to track or recorder to recorder.
6. Mixing of one or more recordings and other signals.
7. Tape cutting and splicing, i.e. re-arranging the order of various items of recorded material.
8. Tape loops (production of continuous rhythmic sound patterns).

The above can be combined in various ways, for example, a recording made at $3\frac{3}{4}$ ips can be replayed at $7\frac{1}{2}$ ips to obtain an increase in pitch and a decrease in the time interval between sounds or musical notes. Artificial echo can be given to signals whilst they are being recorded and the attack/decay properties of a sound can be changed by simple tape cutting. Recording treatment is limited only by the amount and type of equipment (see also Chapter 5).

Sound Sources

Most of the sound sources listed here are available to the amateur and some can be constructed by those with a little experience of home built electronics.

1. Audio signal generators, sine and square-wave.
2. Audio test records containing tones.
3. Simple variable audio frequency oscillators, i.e. multi-vibrator and phase shift oscillators.
4. Noise generator.
5. Sounds recorded via a microphone.
6. Electronic musical instruments.

Control circuits

Quite a number of simple circuits suitable for electronic treatment of generated or recorded sounds are given later but simple treatments can include:

1. Artificial attack and decay.
2. Waveshaping, i.e. filtering, etc.

3. Modulation of one sound by another, such as ring modulation.
4. Reverberation and echo by means of tape head feedback and/or a spring line reverberation unit.

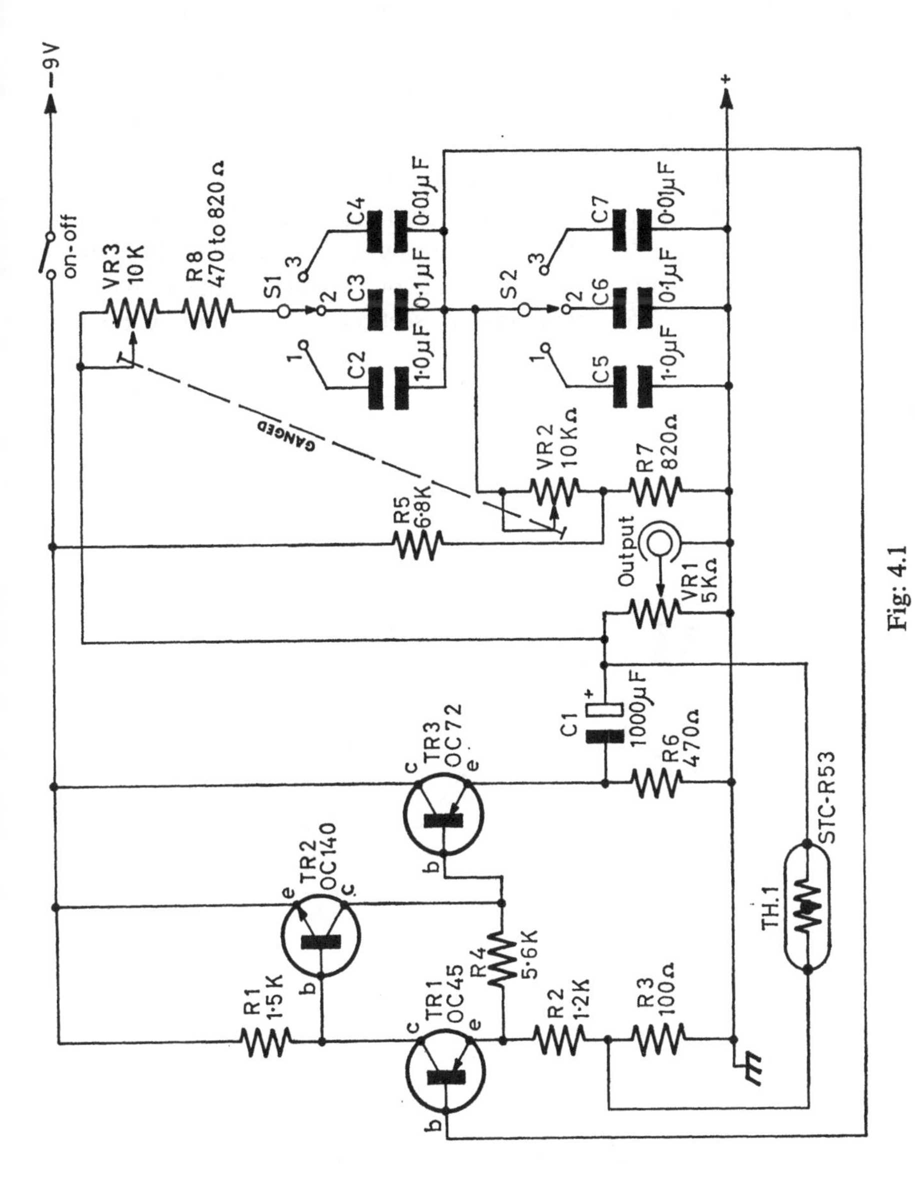

Fig: 4.1
Wien bridge sine-wave oscillator *(By courtesy of Mullard Limited).*

Experimental circuits

The sine-wave generator is one of the most common signal sources for electronic music because a sine-wave provides a pure tone almost completely devoid of harmonics but which can be reshaped by squaring it, i.e. it can be transformed into a square-wave. A circuit for a stable sine-wave generator covering the range 15 to 20,000Hz is given in Fig: 4.1. The circuit is due to Mullard Limited and will operate from a 9V supply. The sine-wave output is approximately 1V rms but can be controlled to any lower level by VR1. Frequency control is in three steps of approximately 15 to 200Hz, 150 to 2000Hz and 1500 to 20,000Hz and is continuously variable over each band. The frequency control VR2/VR3 is a dual ganged potentiometer Colvern type 5018/15F 10Kohms per section and the output stabi-

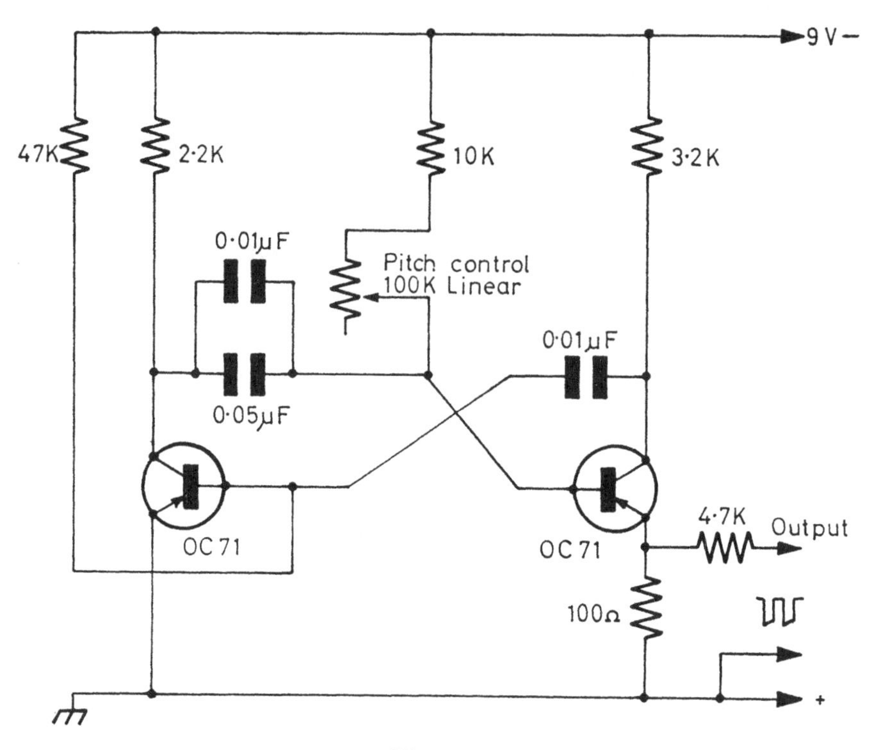

Fig: 4.2
Simple variable frequency square-wave generator.

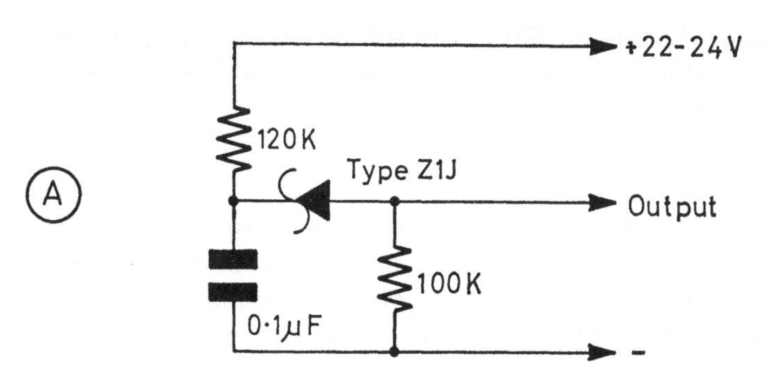

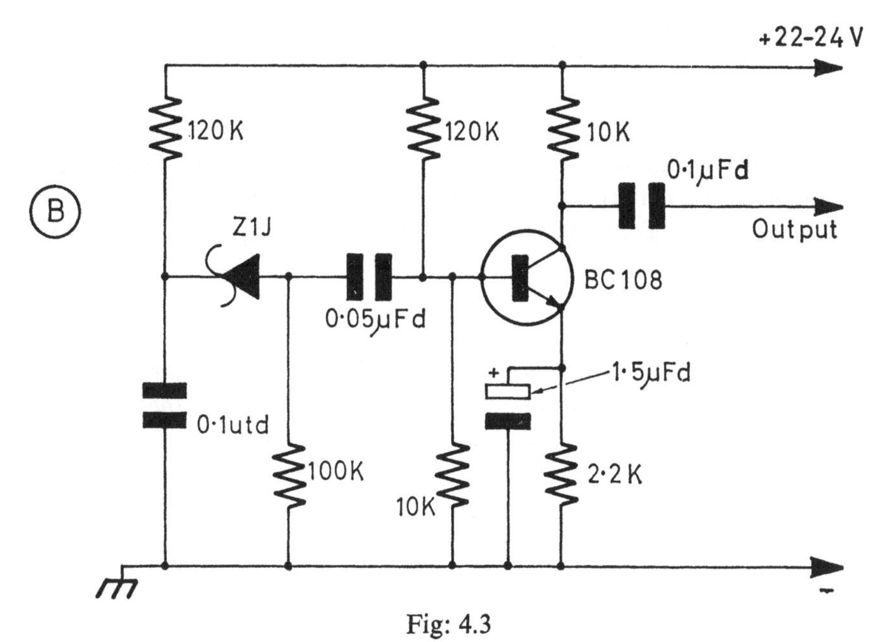

Fig: 4.3
(A) White noise generator (see text references to the noise diode Z1J).
(B) White noise generator as in (A) plus an amplifier.

lizing thermistor TH1 is an STC type R53. The remainder of the components are standard and all should be obtainable from radio component dealers.

Simple square-wave generator

The multi-vibrator circuit shown in Fig: 4.2 will provide a square-wave signal nearly 1V peak-to-peak in amplitude and at continuously variable frequency over at least three octaves beginning at about 1100Hz and going downward to approximately 130Hz.

Noise generator

White noise is a sound that resembles hissing steam and has random frequency, phase and amplitude components. It can be generated quite easily and is another electronic sound source used in electronic music and one which will be found in music synthesizers. A circuit for white noise is given in Fig: 4.3A and is a basic one requiring a special Zenner type diode. So called noisy diodes do not generate sufficient noise signal but a treated Zenner voltage control diode can be made to produce a high noise output covering a wide frequency spectrum. The Zenner diode Z1J used in the circuit, Fig: 4.3A, can be obtained from Semitron Limited, Cricklade, Wiltshire. It will generate between 50 and 100mV of noise which can be further amplified to several volts by using a transistor amplifier as in Fig: 4.3B.

Filters

Filtering and/or waveform shaping is dealt with in Chapter 1. It is not possible to include precise component values for the circuits given as these will depend entirely on the filtering or wave shaping required and on frequency, etc. Circuit values can be determined experimentally, or by reference to information dealing specifically with filter design.

Modulation

There are numerous ways in which one audio signal can be made to modulate another but one of the most frequently used methods is ring modulation. A simple but reasonably effective circuit is shown

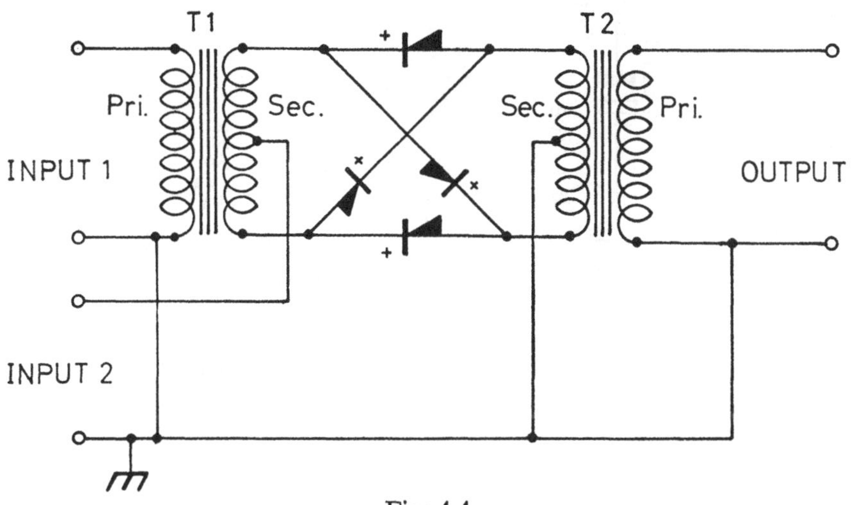

Fig: 4.4
Basic ring modulator

in Fig: 4.4. This consists of two transformers with a four diode rectifier network and it has two inputs and one output. Its function is such that if two tones of different frequency are fed one to each input there will appear at the output a signal consisting of the sum and difference frequencies of each of the two tones. For example if the two tones at the inputs are say 1000 and 1300Hz respectively, the output signal will consist of 1000 + 1300 = 2300Hz and 1300 — 1000 = 300Hz. If the modulator is not accurately balanced the output may consist of the input signals plus the sum and difference frequencies which, for the inputs mentioned above, would be 1300, 1000, 2300 and 300Hz.

Attack and decay control

A basic but click free adjustable circuit providing a controlled degree of attack and decay is given in Fig: 4.5. In this circuit a pulse is generated when the key is closed. The pulse drives the transistor TR1 on and charges C2. The voltage from C2 is used to drive the amplifier TR2 which is normally cut off. The degree of attack can be controlled by varying the value of R which could be a variable resistor. The degree of decay is controlled by the value of C. Signals

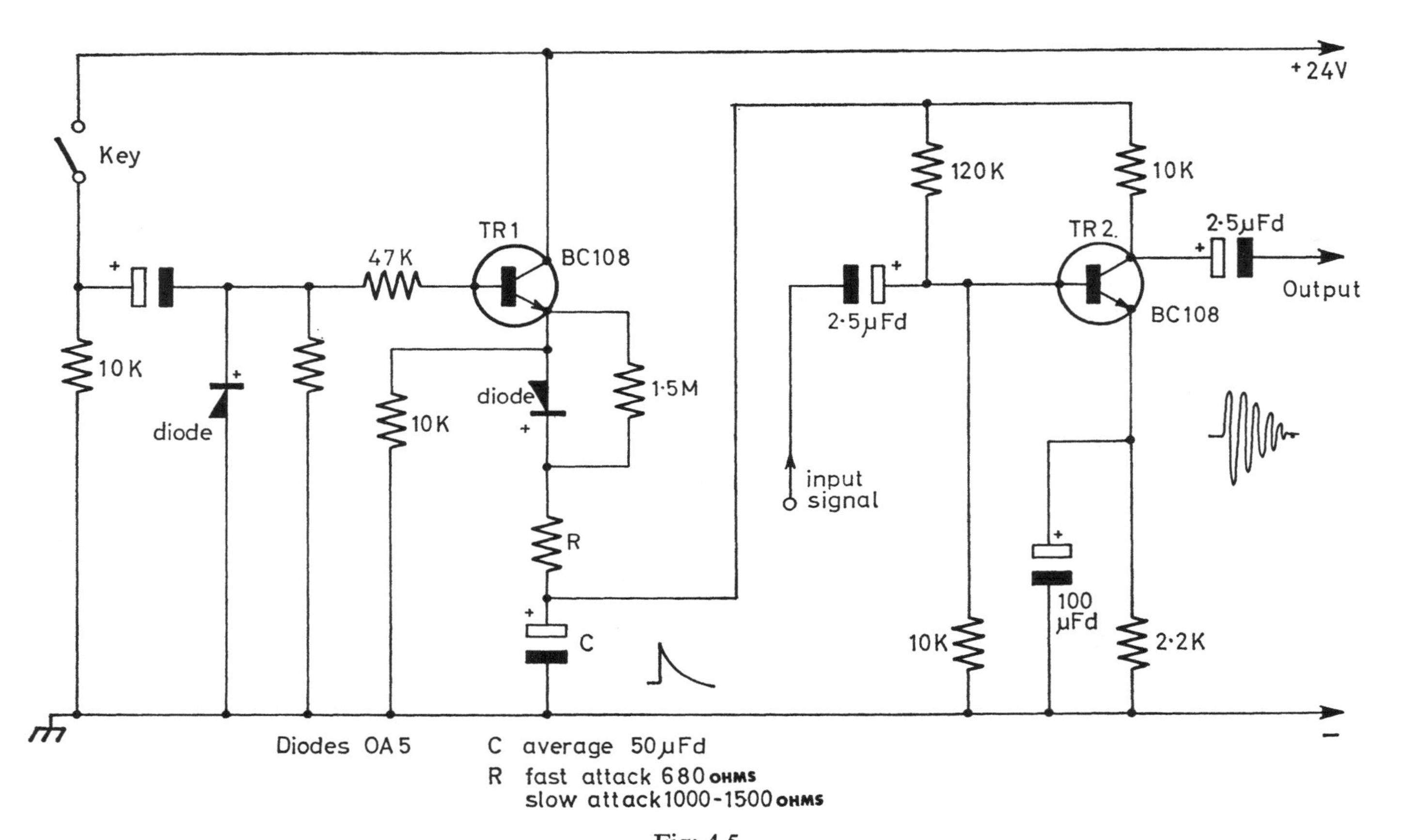

Fig: 4.5

Trigger attack and decay circuit. The signal input to TR2 should not exceed 10mV.

to be controlled are coupled to the amplifier input and can be derived from any continuously running tone generator. The input signal to the amplifier should not exceed a few milli-volts.

Sounds for Science Fiction

The subject of composing electronic music is not within the scope of this book. Knowledge of how to compose electronic music can be obtained only by special training courses available at Universities and schools of music. This need not deter anyone from experimenting or indeed producing musical composition and sound effects, etc., for amateur dramatics, cine films and the like. Indeed the writer makes no claims whatsoever as a music composer but was in fact awarded a first prize for electronic music in a professional recording contest (Scotch Trophy 1965) and at about the same time produced all the special electronic sound effects for a television film series called 'Space Patrol'.

Special sound effects are usually required for science fiction plays and films, etc., but it is first necessary to have a good idea of the story and of what the sounds will be representing. The BBC television series 'Dr Who' for example, required weird voices for beings known as Daleks. They had to speak with intelligible voices, however, which could not therefore be too highly distorted. This was a children's programme. The BBC Radiophonics Workshop (studio for electronic music and effects) simply used a ring modulator with a low frequency sine-wave fed to one input and speech to the other. The Dalek voices therefore consisted of ordinary speech modulated at a very low frequency, generally in the region of 20–30Hz. Imagination is an important tool for this kind of work. Supposing, for example, the sound of a space vehicle take-off is required. Much depends on what type it is and for a rocket ship white noise violently echoed makes a good sound. Something more futuristic might be required and for this undulating low frequency tones produce a good effect. The sound effects possible with electronic tone generators, filters and tape recorders, etc., are practically unlimited but there is no reason why sounds produced by conventional musical instruments should not be used either, which leads now to another but rather mixed form of music known as musique concrête.

Musique Concrête

The name is derived from the fact that real everyday sounds, sounds produced by musical instruments, electronically generated sounds, in fact any recordable sound can be used. Concrête, or real sound, re-shaped to produce this unusual abstract form of music was pioneered by the French composer Pierre Schaeffer. The basis of musique concrête is to take one or more different sounds and re-shape and distort them out of any resemblance to the original and then reconstitute the treated sounds into a planned composition. Most of the techniques for doing this are similar to those used for electronic music. For example, any sound can be given a different pitch by recording it at one tape speed and playing it back at another. The attack or decay of a sound can be removed by simply cutting it from the tape. Any one sound can be recorded, placed into a loop and then replayed at several different speeds so as to produce a 'scale' of pitches. As with electronic music there are no defined rules for composition which leaves an open path for experiment by anyone.

Note: Many of the circuits given in Chapter 2 can be employed for sound generation and control for electronic music and sound effects, etc. The general techniques used for recording cannot be illustrated by photographs or line diagrams. However, further reference to tape recording and particularly to multi-track recording will be found in Chapter 5.

The Electronic Music Makers

Nearly eighteen years ago the people in charge of the Philips research laboratories at Eindhoven in Holland were approached by a music composer. Could he use the electro-acoustic laboratory and all its facilities for producing electronic music? Composer Henk Badings needed tone generators, filters, amplifiers, and above all tape recorders with which to produce his music and was perhaps not surprisingly regarded with some scepticism by the entire staff of the laboratory. What business had a musician in the holy temple of electronics?

As a renowned composer, Henk Badings knew full well the limitations of traditional musical instruments but he knew too that electronics and tape recording could open the door to a new world of musical sounds. His first performance of electronic music for a ballet —Cain and Abel—proved to be the sensation of the Holland Festival of 1956 and in addition he received tremendous ovations in Hanover, Vienna, Monte Carlo and Edinburgh as well. Badings was not alone in his quest for 'new music' for about this time Stockhausen in Germany, Schaeffer in France, Berio and Maderna in Italy, Luennig, Ussachewskey and others in the USA, in fact composers all over the world were looking to the possibilities of electronics in music and soon studios were set up in music schools and even in the electronics and acoustics laboratories of large electronics manufacturers like RCA of America and Philips in Holland. This country was not far behind and two of the best known pioneers are Daphne Oram formerly of the BBC Radophonics Workshop and Tristram Cary, who now both have their own studios. The BBC too realised the possibilities of electronics in music as well as in special effects and accordingly set up a special studio now called the Radiophonics Workshop. Those who have more recently entered the field and who have contributed new ideas and equipment are Peter Zinovieff, one of Britains few experts in computer and music synthesizer techniques and Dr Robert Moog of the USA, designer of the Moog music synthesizers.

BBC Radiophonics Workshop (*Fig: 4.6, following page 96*)

Studios for the production of electronic music look pretty much alike containing as they do vast arrays of electronic equipment and tape recorders. The BBC Radiophonics Workshop is typical and first came into existence in 1957 mainly as a result of certain experimental programmes. It was soon allocated definite service requirements namely the development and production of special effects, mainly electronically generated, for sound and television programmes.

The effects and music created in the workshop are produced mainly by electronic manipulation of all sources of sound including those from sine and square-wave tone generators, electronic musical instruments and music synthesizers. Compositional work in the Radiophonics studios is a deliberate interpretation process which must be motivated by a creative intention. Random selection of contrived sounds combined empirically into a montage does not usually produce an aesthetic result or a composition with artistic merit. Many of the accepted techniques of electronic music are of course used but the aim generally is not to produce music to stand by itself as a separate work. Aside from a few definable compositions some of which have been released on gramophone records, the major work of the studio is to provide special music and sound sequences to form an integral part of a programme in order to accentuate meaning and create dramatic intensity.

The Radiophonic section is composed of two main workshops, a library, staff offices, a technical workshop and engineering office. It is part of the main BBC Maida Vale studio centre and therefore, conveniently adjacent to the five large music studios used for broadcasting. The two workshops each have their own specific equipment consisting of electronic tone generators, filters, keyboard controls, studio tape recorders, music synthesis equipment, and even synthetic sound effects generators. Some of the work created by composers in the BBC Radiophonics workshop is available on gramophone records. One called "BBC Radiophonic Music' contains thirty-one items by various composers and is available from the BBC or record dealers (No. REC25M). It is possible that some of the earlier popular records like 'Time Beat' on Parlophone 45R4901 and the 'Dr Who' theme on Decca F11837, both created in the Radiophonics Workshop, may still be available but these are not sold by the BBC.

Oramics

The Oramic sound system is a music synthesizer devised by Daphne Oram and is basically a method of creating music and sounds from drawn waveshapes and other graphical information. The method has been used in one form or another in several countries and was at one time employed for tone generation in electronic organs in which waveforms were reproduced photographically on a rotating transparent disc. These were then scanned by means of photo cells from which the electrical waveforms were derived.

The Oramics system employs 35mm film on which the waveform information is drawn by hand. Several film tracks are used and can be synchronized with each other whilst they are made to traverse photocell scanners which in turn produce the electrical waveforms. The signals from the sound generating sources are then taken directly to a tape recorder. However, by using a multi-track recorder the sounds produced by each film track can be separately recorded. This provides further manipulation and control before the various parts of a composition are finally mixed to make the master recording.

The RCA Music Synthesizer

An electronic music synthesizer capable of imitating the known voices of musical instruments was first developed in 1954 by Dr Harry Olson of the RCA laboratories at Princeton, New Jersey. An improved model was later built for use by the Columbia and Princeton Universities in connection with research and for the production of electronic music. In this, two keyboard systems were employed to make perforated paper roll instructions for the synthesizer. When the punched paper rolls were re-run to actuate the synthesizer they were also automatically synchronized with two master disc recorders each of which had six recording channels. Four rows of holes in the paper rolls made it possible to select and use sixteen different operations.

The sound generators consisted of twelve fixed frequency tuning fork controlled valve oscillators tuned to the frequencies required for one octave of the equally tempered scale. Two sets of variable frequency oscillators and a noise generator were also part of the synthesizer. The electronics provided for a number of attack, decay and frequency glide characteristics as well as for the filtering necessary for tone shaping, etc. Signal level was also controlled electronically.

The RCA synthesizer might be claimed as the forerunner of the electronic music synthesizers in use today. It was capable of quite lifelike imitations of known musical instruments and indeed some of its 'piano' synthesis was difficult to distinguish from the real thing.

Music from Mathematics

The invention of the computer opened up new possibilities for music composers. Why not couple a computer to a music synthesizer and then programme the computer to do the composing? This in fact is what IBM did and the outcome was a record called 'Music from Mathematics' (Brunswick STA 8523 stereo or LAT8523 mono). The record contains eighteen different items including a computer synthesized voice which sings with synthetic piano accompaniment.

The process of composing music with a computer and sound synthesis equipment is highly complex. Since any sound can be described mathematically, the composer must first determine the numbers for specifying the particular sounds he is interested in. These numbers are then fed into the computer and recorded in the 'memory' circuits. The computer is thus able to store the information from which it can give instructions for an almost limitless range of sounds. Instead of writing a music score, the composer feeds the computer with a second set of instructions for the 'music' he requires or he may allow the computer varying degrees of freedom to select its own musical arrangements at random. From here the computer supplies the required information to actuate the sound synthesis equipment which, in turn, supplies its signals for recording or for reproduction directly from a loudspeaker.

Electronic Music Synthesizers

A name closely associated with modern music synthesizers is Dr Robert Moog who designed and developed the Moog synthesizers but who prior to this produced a number of small electronic musical instruments one of which was a transistorised version of the Theremin. The first Moog synthesizer made a considerable impact in the world of electronic music but was soon followed by others and at the time of writing this book there are several different manufacturers producing synthesizers varying from small models suitable for schools to giant computer controlled systems for studios of electronic music. Even the amateur need no longer be confined to the less expensive techniques of using a few tone sources and the laborious task of cutting and splicing tape because a small portable synthesizer complete with keyboard can be purchased for less than £400.

As already mentioned in this chapter the composers of electronic music soon began to investigate the possibilities of keyboard controlled tone generator systems and even electronic organs were put to use by some. As a result many curious electronic instruments with standard piano type keyboards made an appearance and one example was a so called tone colour organ made by the Miller Organ Company and which is shown in Fig: 4.7, *following page 96*. This could be played as a more or less conventional electronic organ or switched for the production of unusual voicing and incorporated a third manual to provide alternative scaling and other special effects including pitch glide. Another instrument, also employing a keyboard, was the Trautonium designed and built in about 1928. This was a twin keyboard instrument but instead of the usual twelve notes per octave was capable of minute pitch intervals amounting to over 1000 per octave. The small synthesizer shown in Fig: 4.8, *following page 96*, was designed and built by the writer in 1964 and could not only be voiced as required but also made to provide specific degrees of attack and decay. It employed a simplified form of voltage control and incorporated noise generators and vibrato and tremulant generators.

However, the invention of the computer and the building of synthesizers like the RCA version mentioned earlier brought about an integration of techniques from which has evolved the present day instruments. The composers who use music synthesizers and the engineers who design them acknowledge the fact that the generation of complex dynamically varying sounds and the arrangement of these sounds into a complete music composition is now a much more simple task. The secret of these instruments, borrowed from computer technology is 'voltage control' where one or more operating parameters are determined by the precise magnitude of an applied voltage rather than by manual manipulation of panel controls. It is generally easier to electronically change a voltage rapidly and precisely than to re-set a manually operated control. However, in order to take full advantage of 'voltage control', the controlled units that go to make up a music synthesizer must have a fast response and there must be an accurate relationship between the control voltages and the controlled parameters.

Aside from producing musical and percussive sounds a synthesizer can also be 'played' by means of a keyboard (standard piano keys). The required voicing can be pre-selected for imitation of known

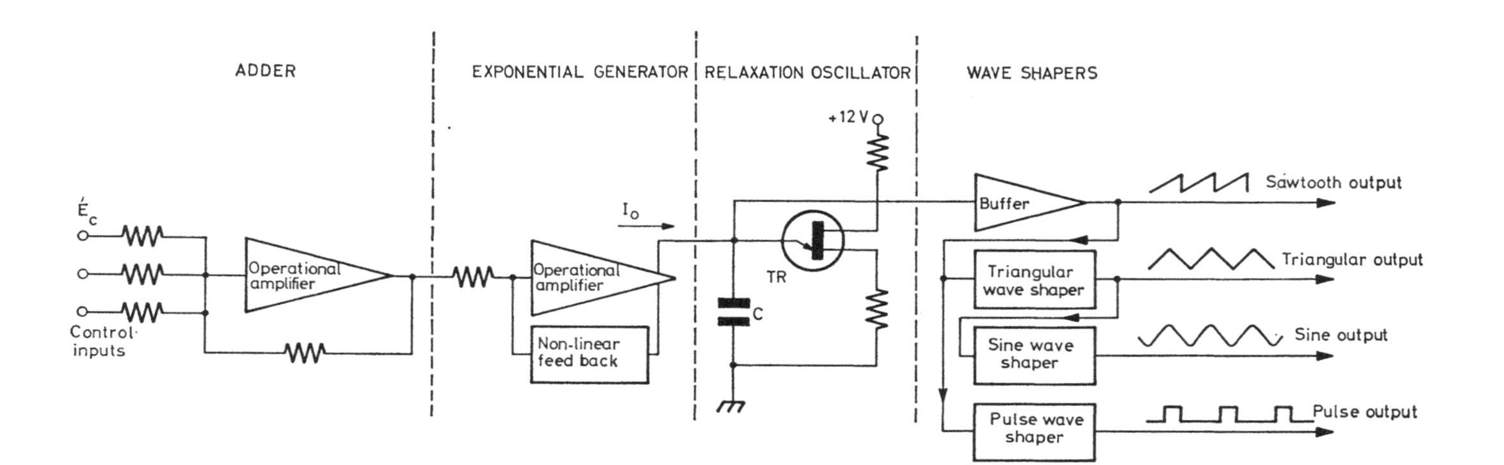

Fig: 4.9
Wide range voltage controlled oscillator (*circuit by courtesy of R. A. Moog*).

musical instruments or those that have never existed, i.e. completely new and hitherto unheard voices can be created. A synthesizer can be programmed to play as an orchestral instrument in 'live' performance but for its more creative applications requires the use of a studio type multi-track tape recorder.

Voltage control

Three important classes of voltage controlled devices are now widely in use and these are oscillators, filters and amplifiers. Voltage controlled oscillators are used mainly to produce audio signals of known frequency and with tonal quality according to waveform. Voltage controlled oscillators (VCO) may themselves be used to control other voltage controlled devices for purposes of modulation, etc. The timing of a musical event may also be achieved by using the output of a slowly oscillating VCO. The diagram in Fig: 4.9 is a schematic of a wide-range VCO with several control inputs so that more than one type of frequency variation may be obtained simultaneously. For example, a slowly varying re-current voltage may be applied to one control input whilst the voltage at another input is stepped in fixed increments. The resultant output would then sound like a musical scale with vibrato (frequency modulation). The control voltages are added and a current I_O proportional to the exponential of the control voltage sum, is derived by two operational amplifiers. These circuits (as in Fig: 4.9) have been derived from analog computer technology. The exponential dependence of the current I_O upon the control voltage sum means that I_O will change by a certain ratio for a given increment in control voltage input change.

In almost all music applications of re-current signals the frequency ratios, rather than absolute frequency differences, are important. In fact a given musical pitch is nothing more than a fixed frequency ratio. For example, an interval of an octave is a frequency ratio of 2:1. The interval of a semitone, the smallest interval in the tempered scale, has a ratio of 1·059:1. Musicians should be able to understand the 'constant' of proportionality between control voltage difference and frequency ratio, for example, a one volt increase in control voltage will increase the frequency by one octave, whereas a voltage increase by one-twelfth will increase the frequency by one semi-tone.

Thus all tones in the tempered scale can be generated by integral numbers of one-twelfth volt increases in the control voltage. It is possible to generate other scales by using different patterns of control voltages.

The output current I_O is used to charge the timing capacitor C and this, in turn, is discharged by the unijunction transistor TR whenever it reaches the breakdown voltage of the unijunction. The resulting voltage across C is a linearly rising sawtooth whose frequency is proportional to I_O and therefore proportional to the exponential sum of the control voltage inputs. With careful design and component selection it is possible to build VCO's whose exponential frequency/control-voltage relationship is musically accurate over a six octave (64:1) frequency range and still useful over a ten octave (1000:1) frequency range.

The sawtooth wave appearing across C is itself useful in synthesizing musical sounds since it contains all the integral harmonics of the fundamental frequency of oscillation. Subsequent filtering, which attentuates some harmonics and accentuates others, is generally used for the production of a wide range of tonal qualities. Three other waveforms which are also musically useful are the sine, triangular and pulse. The sine-wave ideally contains no harmonics but its sound lacks musical satisfaction. The harmonic content of the triangular wave is about 12% and consists entirely of odd harmonics (see Chapter 1) and the sound it produces is somewhat like that of a flute. The harmonic spectrum of a pulse waveform depends on the width of the positive and negative portions. For instance when the positive and negative parts of the wave are equal in width, i.e. as a uniform square-wave with a 1:1 mark-space ratio, then only odd harmonics are present. The sound is musically rich and can be used for synthesis of orchestral tones ranging from the violin to the clarinet.

Voltage Controlled Amplifiers

Amplitude is a most important parameter and is one which determines dynamic range. A voltage controlled amplifier (VCA) capable of varying an audio or control voltage is shown in Fig: 4.10. Like the VCO it incorporates an adder and an exponential generator to process the control inputs. The amplitude controlling elements are

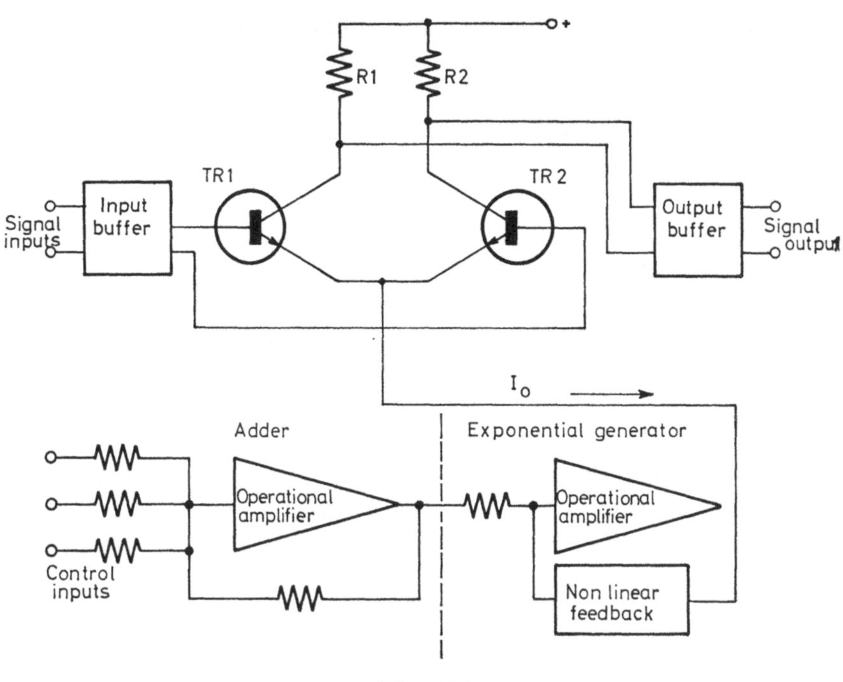

Fig: 4.10
Voltage controlled amplifier (*circuit by courtesy of R. A. Moog*).

TR1 and TR2 which are driven by an input buffer stage with a very low output impedance. Junction transistors have the characteristic that a given base to emitter voltage change will result in a fixed percentage collector current change, regardless of the magnitude of the average collector current. Thus as the standing current in the transistor is increased, the absolute collector current change, for a given base to emitter voltage change, will increase proportionally. The constant base to emitter voltage of the circuit in Fig: 4.10 is assured by the low output impedance of the input buffer stage. The combined standing current through TR1 and TR2 is I_0. The collector current variations appear across R1 and R2, as voltage variations and are amplified in the output buffer stage. Thus the gain from signal input to signal output is proportional to I_0 which, in turn, is proportional to the exponential of the sum of the control input voltages. The gain will therefore increase by a given ratio for a certain incremental increase in the control input voltage. The relation between gain and control voltage is set so that a one volt increase in the control sum will increase the amplifier gain by 12dB. Control characteristics can be accurately maintained over a range of 80dB.

Voltage controlled filters

With a little additional circuitry a voltage controlled amplifier can be used as a voltage controlled filter (VCF). A typical VCF is shown in Fig: 4.11 in which the adder, exponential generator, input buffet and transistors TR1/TR2 are as in Fig: 4.10. The collector currents of TR1/TR2 may be considered as those of the remaining pairs (TR3 to TR10). The inputs to each of these pairs are shunted with fixed capacitors C. At low frequencies the reactance of the capacitor is much higher than the emitter to emitter resistance of the transistor pair so the signal passes up the chain of transistors with little attenuation. At high frequencies the signal is, however, shunted around the emitter to emitter input and is sharply attentuated by the time it emerges from the collectors of TR9/TR10. The circuit therefore behaves as a low pass filter.

The cut-off frequency is that at which the reactance of the capacitor is equal to the emitter to emitter resistance of the transistor pairs. The capacitors are fixed and the input resistances of the transistor pairs are varied by changing the control current Io. A filter of this nature is capable of accurate variations of cut-off frequency over a three decade (1000:1) frequency range. The relationship between the cut-off frequency and the control voltage is exponential and is set to be exactly the same as the relationship between the VCO oscillator frequency and control voltage. A one volt increase in the sum of the control voltage will double the cut-off frequency. The addition of the feedback resistor R_F introduces a narrow resonance peak in the response of the filter at the cut-off frequency and thus converts the VCF from a low pass to a resonant filter. When a sound with noise content, i.e. having many frequency components, is passed through a resonant filter the output signal assumes a pitch, this being close to the resonance frequency of the filter. This allows a composer to work with virtually any sound but which can be given a definable pitch.

Control Voltage Generators

The most important of these are the transient generators for producing attack and decay and are of great value in producing rapid changes in frequency, waveform or amplitude. It may be required for example to start off a sound with low harmonic content and this is produced by applying a rising transient control voltage to a VCF so that the filter first passes only the fundamental of a waveform and then later the harmonics as well. On the other hand a sound beginning with a high harmonic content may be required followed by rapid reduction of the harmonics. A falling transient voltage is used in this case in conjunction with a voltage controlled filter.

Periodic control voltages are also useful in imparting frequency modulation such as vibrato, tremulant and tremolo, etc. Random control voltages can be derived from white noise and are often used to introduce a degree of uncertainty to any of the voltage controlled parameters thus producing an aural interest to an otherwise steady tone. Specialized function generators such as staircase waveform generators are also used to create distinctive patterns.

The total electronics of any music synthesizer at present available, large or small, is far too complex to deal with entirely in this book. The basic principles of voltage control outlined in the previous paragraphs and by kind permission of R. A. Moog Inc., are employed in Moog synthesizers and are similar to those used in other makes. The music student would gain little from detailed descriptions of synthesizer circuitry, unless he happened to be an electronics engineer and the amateur electronics enthusiast would be unable to build himself a synthesizer, even if provided with the circuits, because many of the special components and integrated semi-conductor devices would be unavailable. However, it must be mentioned here that at least one or two firms manufacture and supply ready to use voltage controlled modules, which includes oscillators, filters and amplifiers, etc., and it is from these that the construction of a music synthesizer is possible.

For those who have a particular interest in this field the rest of this chapter is devoted to some of the music synthesizers at present on the market. It must be emphasized, however, that development is very rapid and that even by the time this book appears in print new and much more sophisticated models may have become available.

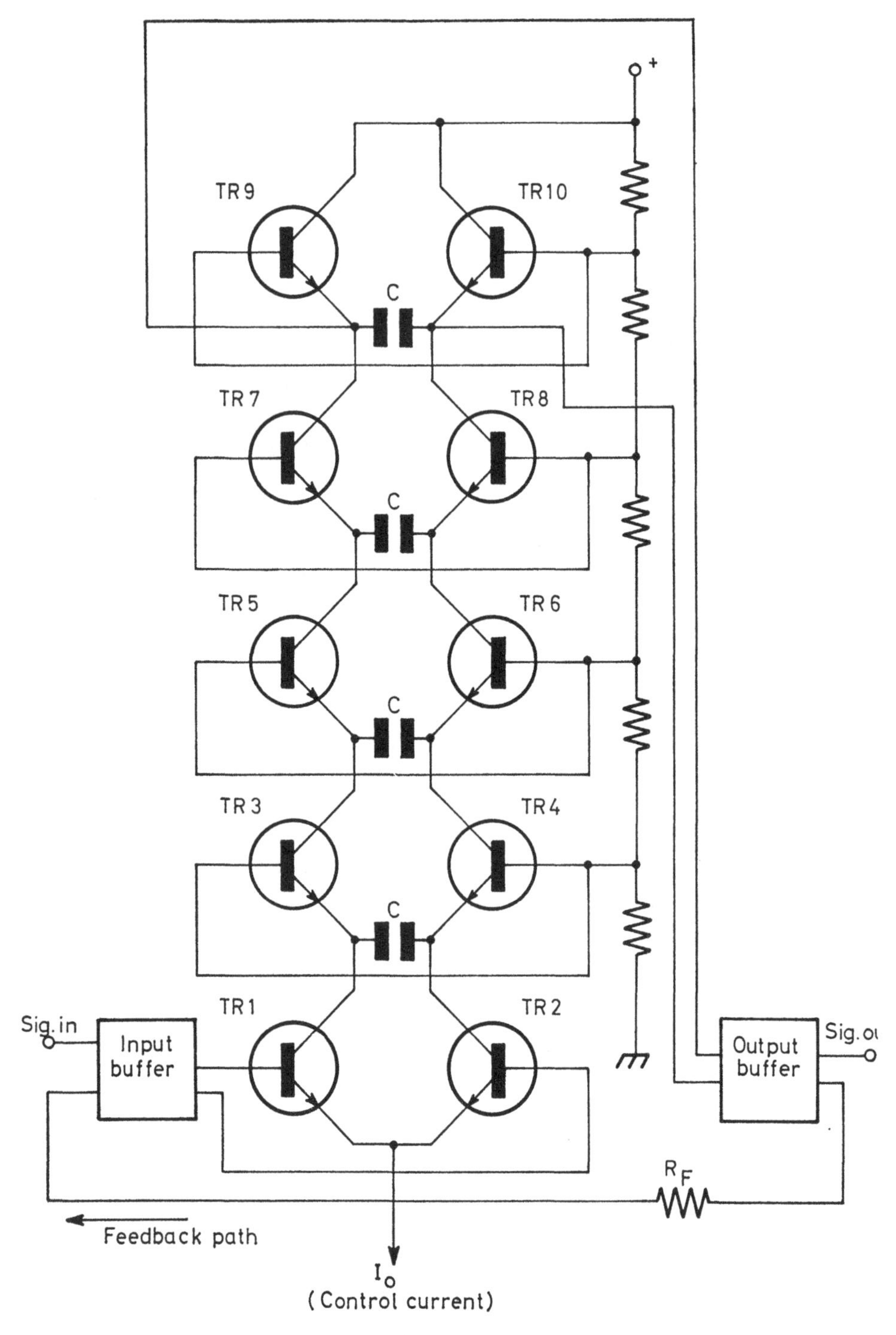

Fig: 4.11
Voltage controlled filter (*circuit by courtesy of R. A. Moog*).

Moog Synthesizers

At least eight different models are available plus a range of accessories including sequential controllers, studio signal mixers, keyboards and even separate modules from which to build up a complete synthesizer. One of the most recent additions to the Moog range is their 'Mini-Moog' intended for small studios, and schools, etc. This incorporates three VCO's, a noise source, filters and keyboard controller. It can be used to process external audio signals and can be used 'in concert'. Unlike most synthesizers it does not use patch panels so all operations are switch selected. All the Moog synthesizers now include keyboard control and the various other larger models are designed mainly according to requirements. The largest is their MK3C shown in Fig: 4.12, *following page 96*, which is intended for large institutional and commercial electronic music studios. This has a single five octave keyboard and incorporates a reverberation unit, a four channel mixer, nine oscillators, three voltage controlled amplifiers, two filters and a host of other voltage control devices. Some of the special features of Moog synthesizers are that all modules are designed for the widest possible range of applications and voltage levels and adjusted so that logical interconnections between instruments may be set up with a minimum of patch cords and without level adjustments. Moog synthesizers are completely compatible with standard professional studio equipment and can all be used in 'live' or concert performance.

EMS Synthesizers

Some idea of the requirements for a small electronic music studio built up around an EMS Synthi music synthesizer can be obtained from the diagrams in Fig: 4.13. The Synthi is a small portable instrument with a wide variety of applications, including live performance by itself, or in conjunction with other musical instruments. It can be used as a teaching aid to demonstrate acoustic phenomena and can be keyboard controlled. The external equipment as shown in Fig: 4.13 may be called 'peripherals' and includes tape recorders, amplifiers and loudspeakers, etc., as well as extra devices such as the Synthi Sequencer which enables sequences of control voltages to be stored and replayed. Other 'extras' may include an octave filter bank and a

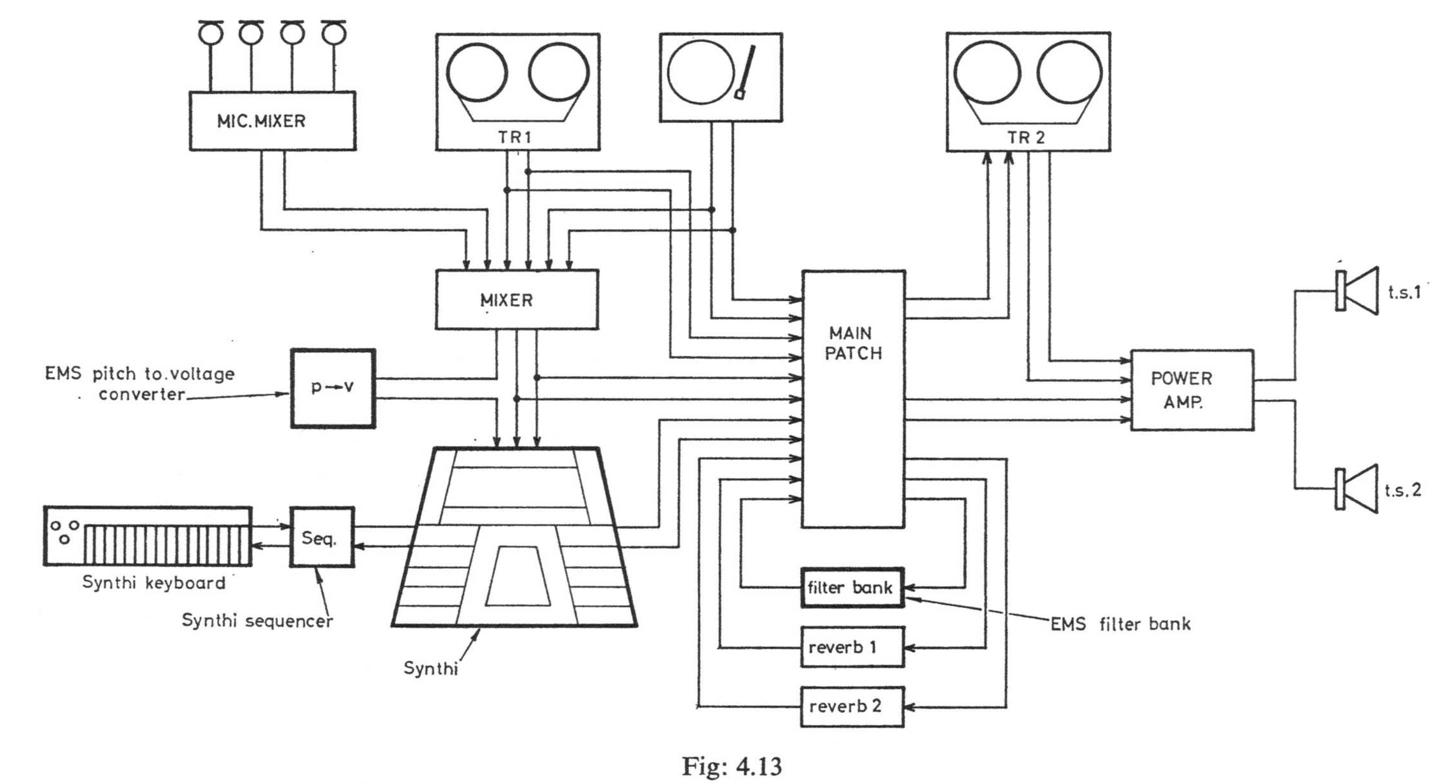

Fig: 4.13
Typical small electronic music studio layout around an EMS Synthi music synthesizer.

Fig: 4.15 The 'Musys' studio of Peter Zinovieff at Putney in London. To the right can be seen the large EMS Synthi 100 computer controlled music synthesizer.

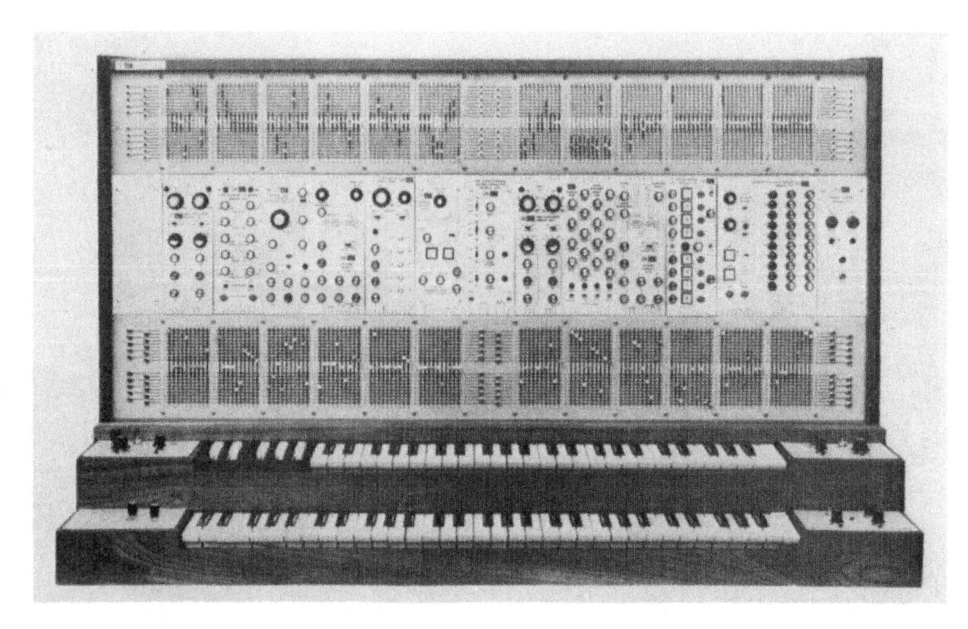

Fig: 4.17 The Tonus ARP 2000 features the matrix switch patching system and has two control keyboards.

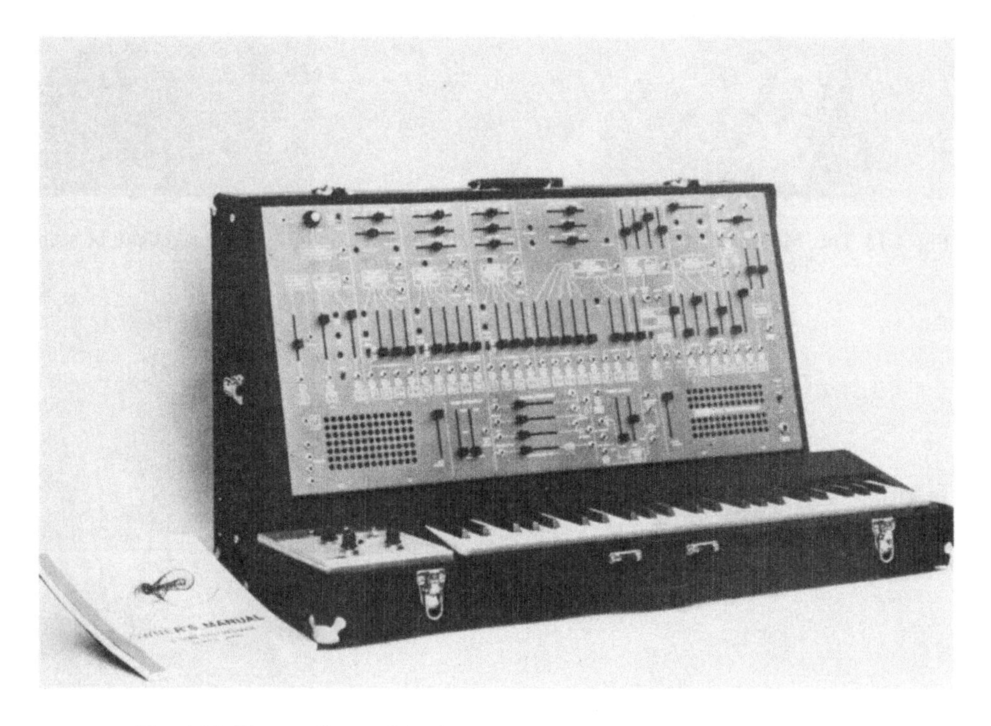

Fig: 4.18 The small portable Tonus ARP music synthesizer Model 2600.

Fig: 5.1 The Studer A80 16 channel professional studio tape recorder.

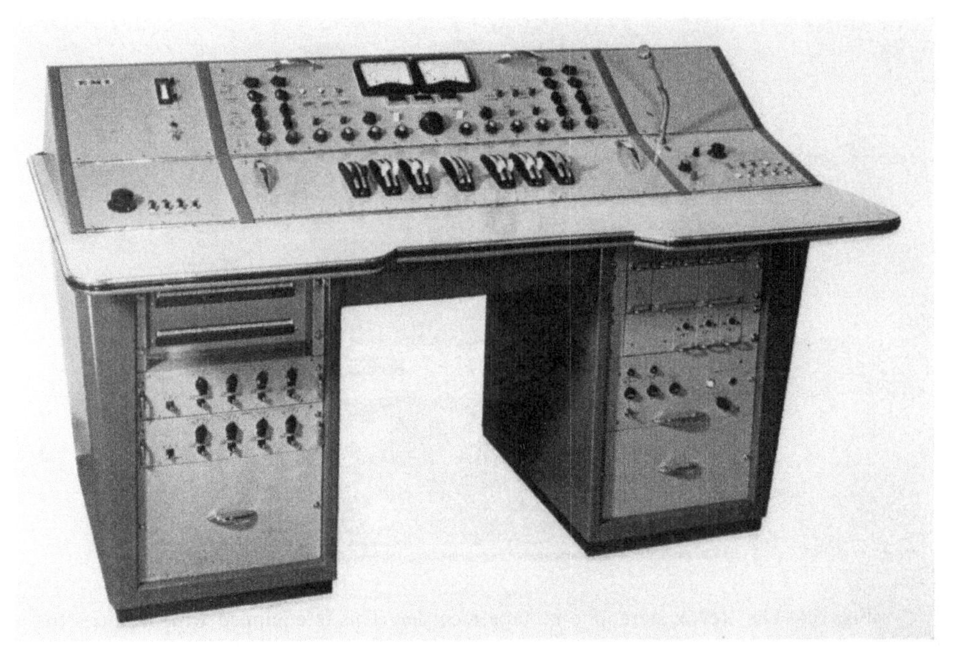

Fig 5.2 A stereophonic, two channel signal mixing consol by EMI Limited.

Fig: 5.3 A Dolby B type noise reducing unit. This model by TEAC can be used with any tape recorder and will reduce inherent noise by up to 10dB.

Fig: 5.6 The Revox stereophonic tape recorder. This is equipped with facilities for track-to-track recording, mixing and tape echo.

pitch-to-voltage convertor. One of the special features of the Synthi is a pin connected patch board which involves simply placing pins in holes in the board to obtain instant connection between one device and another. This can be seen clearly on the front of the EMS Synthi A model shown in Fig: 4.14, *following page 96*, similar to the Synthi but is an otherwise completely portable instrument. The EMS range also includes giant computer controlled synthesizers such as the Synthi 100 which can be seen in Fig: 4.15, *following page 128*. This photo also shows the whole studio for electronic music belonging to Peter Zinovieff who has offered to give it to the nation for the benefit of students and young composers providing someone will bear the cost of re-housing and maintaining it. The equipment is valued at around £40,000. It would make the nucleus of an electronic music centre not only for this country, we have none at present, but also for Europe as well. As Tristram Cary pointed out in an article in *Studio Sound* magazine, the development of electronic music in Britain has been mainly due to a handful of pioneers running private studios. It is essential that this country should have a centre of its own. In order to create a body of support for a 'National Studio' a *British Society for Electronic Music* has been founded (1969) and those who are interested should write for details of membership to BSEM, 49 Deodar Road, London, SW15.

All EMS equipment is the result of teamwork design by Peter Zinovieff, Tristram Cary and electronics engineer David Cockerell. The EMS range are the only British made synthesizers at present on the market and are used by the BBC Radiophonics Workshop as well as by large recording studios and universities, etc.

The Tonus ARP Synthesizers

The profusely illustrated operational manuals issued with all ARP music synthesizers take the owner very gradually from basic sound generation right through to the complex possibilities of the exclusive ARP matrix switch patching system which links all the functions of the synthesizer within itself or with external sources, e.g. amplifiers and loudspeakers for live performance, recording equipment for multi-tracking, etc., or to other signal sources which may be musical instruments. Within the ARP synthesizers is of course the now more

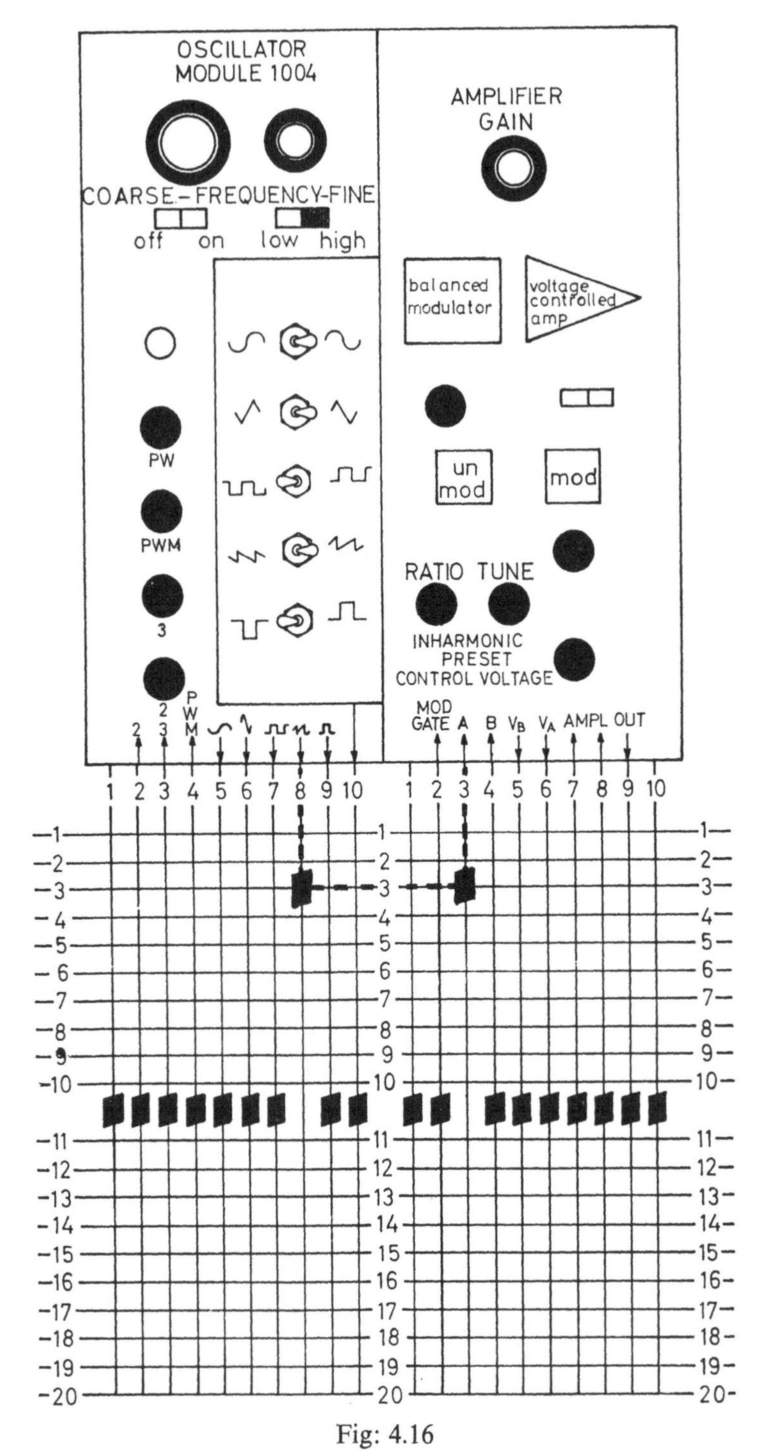

Fig: 4.16
The Tonus ARP Matrix switch patching system used on their larger synthesizers.

or less conventional range of voltage controlled oscillators, amplifiers and filters, noise generators and ring modulators, etc., etc., but the method of coupling these is quite unique. The ARP matrix switch patching system enables the operator to connect any one input or output to any other input or output on any of the modules or to external sources. Some idea of the simplicity of the matrix switch system can be gained from the diagram in Fig: 4.16. The small solid blocks represent the switch heads. These can be moved up or down or from left to right and in the example given the output from an oscillator on vertical line 8 is connected through to an amplifier input via vertical line 3. The link across is made via horizontal line 3 but, had this been occupied, then any other horizontal line could have been used. The permutations possible with this form of linkage are enormous but do allow the operator to make the connections quickly and accurately. All the larger ARP synthesizers like the model 2000 shown in Fig: 4.17, *following page 128*, employ the matrix switch patching system. The smaller ARP portables like the 2600 model shown in Fig: 4.18, *following page 128*, do not have a matrix switch patching system but instead employ a plug and cord linkage. These portables are nevertheless flexible in use and can be used in 'live' performance by themselves or in conjunction with other musical instruments. The Tonus ARP equipment is of American manufacture but is available in the UK (see references to suppliers, etc., at the end of book). They can also supply voltage controlled oscillators, filters, amplifiers and other units for synthesizer construction.

CHAPTER FIVE

The Tape Recorder in Music

The finest known medium for sound and music recording is magnetic tape. All music intended for distribution on gramophone disc records is first recorded on magnetic tape, many broadcast music programmes come from tape recordings and of course the vast majority of television programmes are pre-recorded on tape. Electronics play a major part in this, in fact without them magnetic tape recording would not be possible. Aside from tape recording equipment the modern professional recording studio also employs a considerable array of highly complex auxilliary electronic audio devices such as sound mixing units, noise reduction systems, signal limiters, reverberation units and monitoring equipment, etc. The total cost of a studio, including recorders, can run into many thousands of pounds. Magnetic tape recording has, however, become very domesticated and worthwhile equipment for home use can cost less than £200. For this amount one can obtain a reasonably good stereo tape recorder complete with a microphone and make recordings of sounds and music with quality very nearly equal to that obtainable from a good LP record. This does not mean that the amateur, even using a top quality domestic or semi-professional tape recorder, will be able to obtain the perfection in balance and acoustics that is possible with studio equipment but with care, the results need not be far short of this. Most of the satisfaction stems from achieving the best that knowledge and equipment will allow.

A few words about professional studio equipment may be of interest if only to show just how sophisticated and complex it can be. A modern studio that has to cater for all kinds of music recording and particularly modern 'pop' will most likely employ one or more recorders like the sixteen channel Studer machine shown in Fig: 5.1, *following page 128*. This is one of several versions of the Studer A80 models which begin with a two channel model using standard quarter-inch wide tape and progress through four and eight channel models which use half-inch and one-inch wide tapes. The sixteen channel model uses two-inch wide tape. Tape recorders of this nature are precision engineered and cost several thousand pounds. Considerable emphasis is placed on a performance, that must not only be to a very high standard but must be maintained throughout years of operation for long periods. Nearly all studio recorders of this nature operate at 7½ or 15 ips and have a uniform frequency response of averagely 30 to 18,000Hz. Signal to noise performance is usually better than −60dB and wow and flutter (cyclic departure from nominal tape speed) is normally less than 0·1 %. The multiple track machines allow recording on any one or more channels simultaneously and then further recording on one or more channels in synchronism with tracks already recorded. This is a technique frequently adopted in studios whereby certain parts of music are recorded and others are added later. The multiple track system is also used for stereophonic recording when it may be required to later alter the spatial relationship of various orchestral instruments, i.e. to have them sounding from a position (during reproduction) different from that occupied in the studio. Solo parts are sometimes recorded later on a separate track when it has not been convenient or possible for acoustic or other reasons to record the solo part with an orchestral accompaniment.

Auxilliary equipment

Perhaps the most essential piece of studio equipment, second to the tape recorder, is the signal mixing unit which may have to accept the outputs from several microphones and other signal sources, possibly even from some of the replay channels of a multi-track recorder, and blend or divide them appropriately for recording. The mixing unit may have to route signals through tone correction networks or through reverberation units and must also be able to provide a composite signal, i.e. that being recorded, to a loudspeaker monitor for the recording engineer and producer to hear what the complete recording will sound like. In addition, the mixer must show the levels of all signals being mixed because electrical signal balance is just as important as aural balance. Mixer units range from small stereo consoles like that shown in Fig: 5.2, *following page 128*, to enormous multi-channel systems designed to match with multi-track recorders like the Studer A80 shown in Fig: 5.1, *following page 128*.

Reverberation units play a great part in sound recording for with these the 'acoustics' can be added to music recorded in an otherwise acoustically dead studio, i.e. one that has sound absorbing materials on the walls to prevent reverberation. The amount of reverberation required may be that produced by a small room to the highly sustained reverberation of a large cathedral. For this purpose metal plate reverberation units are normally employed which are massive devices with two or three plates around six by four feet and which are housed in a vibration free and sound-proof container or room. Spring line units may sometimes be used by a large studio for single circuits and are favoured by small studios for general use because of their lower cost (see Chapter 3). Tape echo units are also frequently employed for special effects and so also are signal compressors and similar devices which are used mainly for controlling dynamic range. Tone control networks, more frequently called 'curve benders' are rather more complex than the simple bass and treble control networks used in domestic hi-fi equipment. Studio curve benders allow an incremental increase or decrease in response over narrow band widths within the full audio frequency spectrum, thus permitting frequency response cut or lift at bass, lower midrange, true midrange, lower treble and upper treble. Sometimes as many as eight narrow bands are utilized.

A very recently developed and now much used device is the noise reducer. Those at present in use are due to Dr R. Dolby, the designer and therefore usually referred to as Dolby units. One of the greatest problems in sound recording is keeping random noise to a minimum. Amplifiers and tape recorders produce a certain amount of noise inherently but which will build up when successive amplification and recording processes are used. With a first class studio tape recorder the noise level from a master recording may be very little and in terms of rms voltage some 1000 times less than full recording level, i.e. −60dB. During very quiet passages of music even this noise level would be just audible. However, successive recording, which may include copying the master tape to a master disc for record pressing or a primary tape for multiple copying of secondary tapes, can introduce more noise and easily result in a noise level of −50dB or only 316 times less than the sound level. Replay of the copy discs or tapes by the user may introduce still more noise resulting in a possible noise level of −40dB or only 100 times less than the sound level. In this case the noise level would certainly be audible even to the point of annoyance. Unfortunately noise is composed of random frequency, amplitude and phase, and cannot be filtered out with conventional frequency filters. The Dolby method of reducing noise is essentially one that suppresses the noise by first raising the level of weak signals (quiet passages in the music) which can be done during recording. The noise level due to the recorder remains constant. However, on playback what were the weak signals are now restored to their original level. This means that the noise which was recorded at its normal level is automatically reduced as the signals are reduced. The improvement in signal to noise can be as much as 10 to 15dB. A recorder having a normal signal to noise performance of −60dB (1000 times less than the signal) can now be made to produce recordings with a noise level of −70 to −75dB or 3162 to 5623 times less than the signal. The Dolby process is a two part one whereby one part of the process is carried out during recording and the other part is applied during playback. The system has a distortion factor of 0·1% or less and does not effect frequency response. It is now being applied to domestic recording and reproducing equipment and a unit similar to that shown in Fig: 5.3, *following page 128*, can be purchased for use with a domestic tape recorder. Another offshoot of the Dolby system is that pre-recorded music tapes can be pre-Dolbyized, i.e. given the first part of the process and then replayed by domestic equipment fitted with the necessary unit for completing the other half of the process.

Music Applications

Aside from its normal use for recording orchestral and vocal works, etc., in a studio, a tape recorder can be used creatively for the production of electronic music employing the techniques outlined in the first part of Chapter 4, or with music synthesizers. It can also be used successfully by amateur musicians for 'multi-track' music in which one musician plays all the parts for an orchestration with one or more musical instruments. Further, a tape recorder is also a valuable aid in music practice.

It is not really necessary to understand how a tape recorder works but for those who are interested references to books dealing specifically with this subject will be found in the Appendix. It is more important that the owner of a tape recorder and any auxilliary equipment that goes with it, knows precisely its capabilities and moreover its limitations in terms of recording and reproductive quality. A few words about tape recorders and their facilities and about auxilliary equipment may, however, prove useful, particularly for those who do not already own a tape recorder.

Tape recorders more or less begin with small portable cassette machines which are useful for playing pre-recorded music cassette tapes and for recording sounds outdoors, etc., as many are battery operated. Cassette tapes are those of narrow width (about one-eighth inch) wound on a small spool enclosed in a plastic container. The tape is wound onto another spool during recording or playing. The plastic containers are usually sealed which makes it difficult to cut the tape for editing, etc. In any event the facilities of these small recorders are quite inadequate for creative forms of recording where it may be required to cross-track record, provide echo and utilize a mixing system when dealing with one or more sound sources. There is another similar type of recorder that also employs an enclosed tape and this is the 'cartridge' machine. A tape cartridge is similar to a tape cassette in that the tape is completely enclosed except that in this case a single spool is used. The tape winds off from the outer layer, goes across the tape head and returns to be re-wound on the inside of the spool. The whole tape is in effect an endless loop. However, in the case of cartridges, also used mainly for pre-recorded music, the tape is standard width, i.e. quarter-inch wide. Again this system is not suitable for creative recordings and for this we must turn to conventional spool to spool recorders, i.e. recorders on which the tape winds from one spool, across the tape heads and then winds up again on a separate spool. Neither of the spools are enclosed.

Professional spool to spool recorders normally use standard

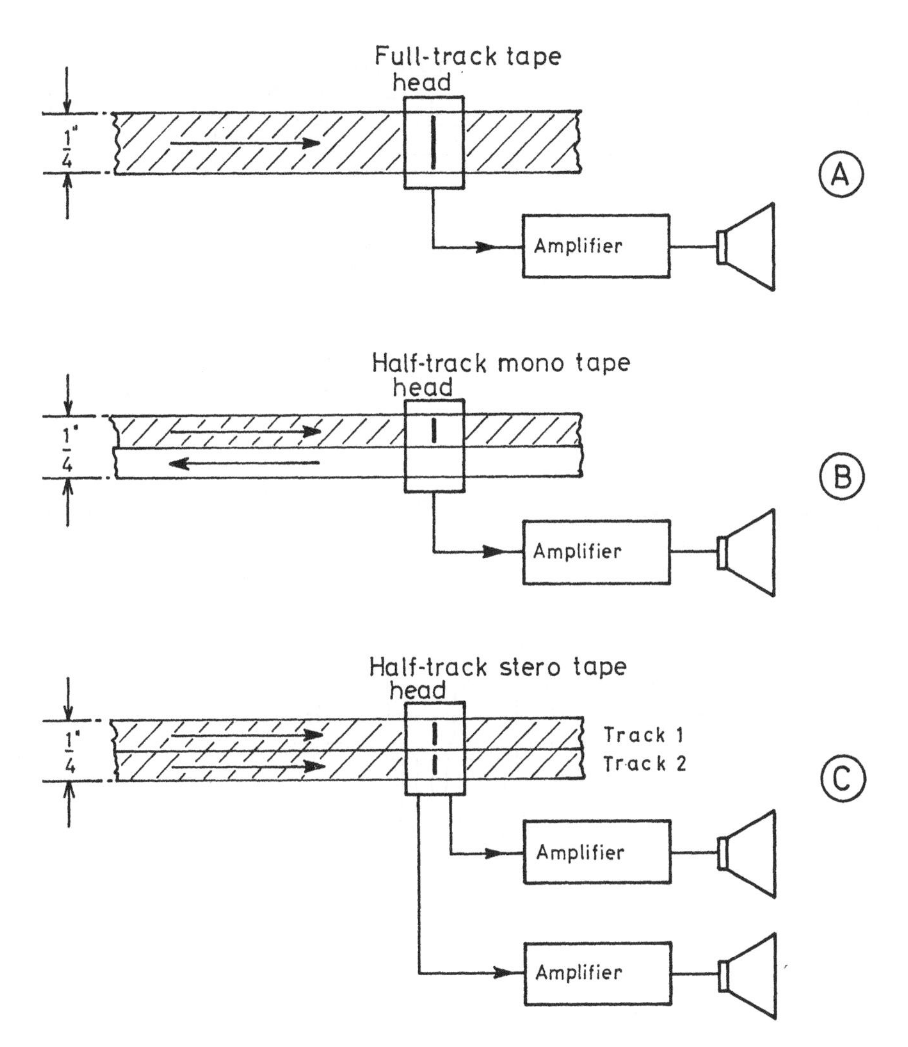

Fig: 5.4

(A) Full-track recording with standard quarter-inch wide tape. (B) Half-track (mono) recording with quarter-inch wide tape. (C) Half-track (stereo) recording with quarter-inch wide tape.

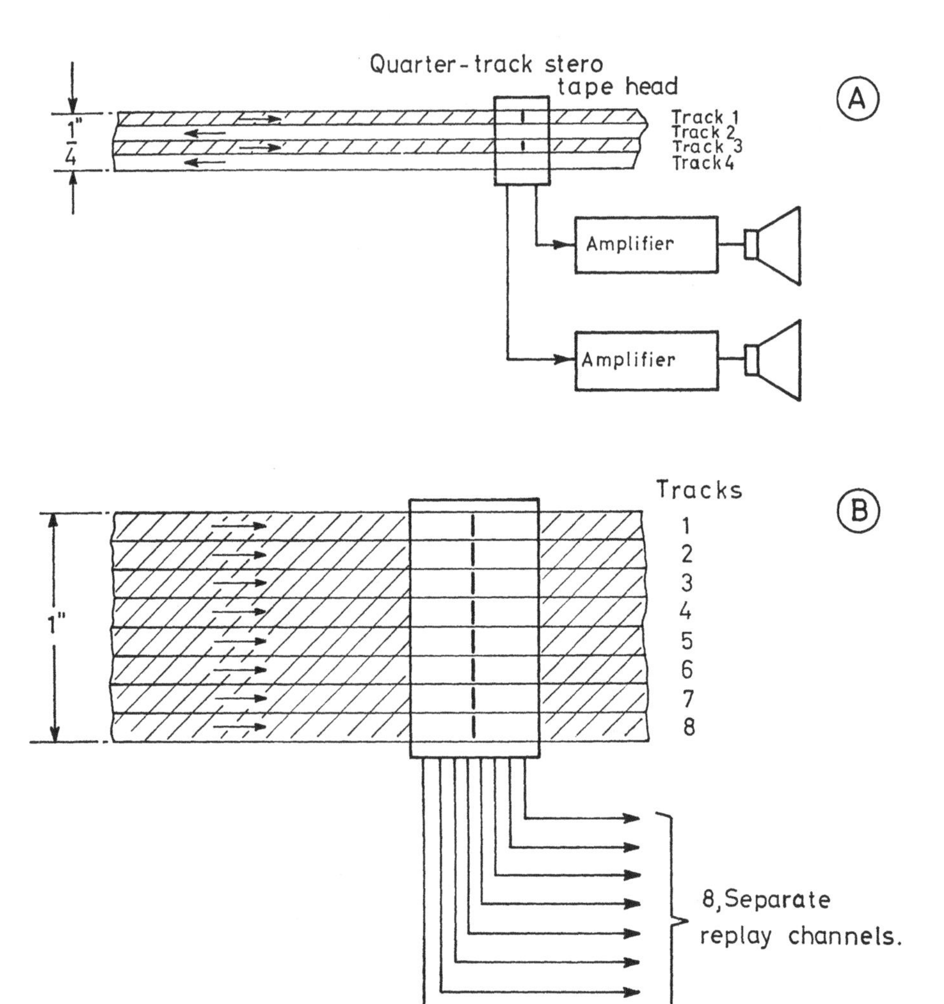

Fig: 5.5

(A) Quarter-track stereo recording with quarter-inch wide tape. Each track can be used for mono recording. (B) Multiple track recording system used on professional studio recorders. Eight separate tracks on one-inch wide tape.

quarter-inch wide tape and can record monophonic, using the whole width of the tape, or stereo, in which case half the width is used for each track because stereo requires two separate channels (see Fig: 5.4). This division of tape width goes further with domestic spool to spool recorders in that up to four separate tracks may be used each becoming a 'quarter-track' (one quarter of the total width of the tape) in which case two separate pairs for stereo can be accommodated within the width of standard tape (see Fig: 5.5A). Domestic tape recorders are readily available for what is called half-track stereo or mono (track division by two) or quarter-track stereo or mono (track division by four). Professional tape recorders rarely, if ever, use less than half-track and the trend now is to use a much wider tape, either half inch or one inch wide, when multiple tracks are required. For example, a one inch wide tape can be divided into four quarter-inch wide tracks or eight one-eighth-inch tracks as shown in Fig: 5.5B. Domestic tape recorders use only quarter-inch wide tape so track widths are always either one-eighth-inch (half-track) or one-sixteenth-inch (quarter-track). The track widths are in actual fact slightly less than those given in order to include 'guard bands', i.e. a small space between each track to prevent the recording on one track overlapping another due to minute differences in alignment between recording and replay heads or due to tape alignment.

Tape Recorder Facilities

For purposes of creative recording a half-track tape recorder is recommended because the overall performance with regard to frequency response and signal to noise performance is better than that obtained with quarter-track. This is very important in track-to-track recording between one tape recorder and another, or from one track to another on the same tape recorder.

Most domestic tape recorders are stereo/mono, half-track or quarter-track but not all have proper facilities for track-to-track recording. Those who are interested in this particular recording technique should make sure that a tape recorder purchased for this purpose has the right sort of facilities. A machine suitable for track-to-track (sometimes called sound-on-sound) should have separate recording and replay heads and each recording and replay channel should be capable of being operated independently. Such a recorder is shown in Fig: 5.6, *following page 128*. Some recorders have built-in switching for track-to-track work but have no provision for mixing and balancing, i.e. no independent control over the signals being re-recorded and those to be recorded with them. The ideal system is one in which each channel of the recorder can be operated independently of the other, i.e. in which one channel can be set for recording whilst the other is actually playing back. In this case an external signal mixing unit can be used. The signals being played back are taken into the mixer together with the signals that are to be recorded with them. The output from the mixer is then taken to the channel that is recording. In this way all signals being recorded are in synchronism which is most important for multi-music recording. It would be as well at this point to explain that multi-track music is the production of an orchestrated piece of music all the parts being played by one person using the same or a different musical instrument for each part. This technique need not be confined to conventional music but can be applied to electronic music, sound montage and even plays where all speaking parts are performed by one person using different voices and which may include sound effects. Before dealing with the techniques of multi-track music, however, a few of the ways in which a tape recorder can be used for music practice may prove useful to those learning to play musical instruments and may also provide some initial understanding of the musical aspect of multi-track work.

Music Practice with a Tape Recorder

There are of course many 'aids' for learning to play musical instruments. Electronic organ manufacturers offer a good deal in this direction. For example there are simplified instructional visual aids with special colour diagrams, keyboards on which the notes for different chords can be illuminated, gramophone and tape records of exercises, accompaniments and rhythm parts against which to practise and more recently, built-in cassette tape recorders (see Chapter 2) on which recorded instruction can be played or which can be used to record one's own playing.

Properly used the tape recorder can be a very valuable aid to music learning and practice. It cannot replace the qualified teacher but aside from its many other possibilities which will be dealt with later, it will allow you to hear yourself play as others hear you—mistakes and all! The tape recorder can be a ruthless unrelenting master when it comes to pointing out errors in playing but it will not correct them for you. Only you can do that. However, a music teacher or someone who is a competent player should be the final judge of your progress, for it is possible to repeatedly make the same mistakes without being aware of them and still not hear them when you listen to a tape recording of yourself making them.

Those who are teaching themselves to play a musical instrument have to rely a good deal on the criticism or appraisal of others who may not be sufficiently qualified anyway. Here the tape recorder can be used to great advantage but only if you carefully follow the recording of your playing with the written score. Count every beat and listen intently for the time values of notes. Listen to the rhythm, the interpretation, the chords, everything. Replay the tape several times, spot the errors, practise the correction of these and make another recording. Don't erase the one with the mistakes so that you can compare it with the one containing your corrections.

Aside from using a tape recorder to hear your own progress there are numerous ways in which it can be used to master difficult phrases and improve timing and rhythm in playing. The more you progress the more the tape recorder will assist. It can be used as another 'musician' to enable you to play duets, i.e. you can first record the basic accompaniment and then play the harmonies and/or melody with it and it can also be used as a 'metronome' to play a given number of bars or to play pre-recorded drum rhythms. As your playing ability and technique progresses the tape recorder will reveal your personal style, which you may not be aware of, unless you listen very frequently to recordings of yourself playing.

Let us now look at the practical ways in which the tape recorder can help, for instance in counting and timing. A metronome is very useful for keeping a steady rhythm but it cannot stop of its own accord at the end of say, 32 bars, thus indicating that you have come to the end of a common 32 bar score. Whilst playing against a metronome it is quite possible to miss a beat or two or even play a few extra but still be in time with the metronome. You could therefore quite happily come to the end of a 32 bar score having used up more, or less, than the 32 bars. In order to make a metronome sound out a precise number of bars, all you have to do is to record its sound for the requisite number of bars. This can be done by releasing the tape recorder pause button a fraction before a beat and then count off the number of beats required. In the case of 32 bars this would be 128 beats. A fraction of a second before beat 129, i.e. immediately after beat 128, stop the tape. On replay you should have exactly 128 beats. But this is not sufficient as you will not know exactly when the first beat is going to occur, so you will require a few extra beats to 'count in'. Four beats will be sufficient making 132 in all. To start the music on beat one simply count one, two three and four and commence on the next or first beat proper. Providing no beats are lost or gained during playing you should finish on the last beat recorded. With a little practice it is possible to record various combinations such as 32 bars plus a 4 bar introduction and 2 bar ending, etc., etc. If you do not have a metronome any rhythmic tapping sound, such as a pencil on a tin box will suffice, providing the rhythm is perfectly even. By using the tape editing technique sets of bars can be produced by cutting the tape at the end of the number of beats required in which case record more than the total number required and cut the tape immediately after the last required beat. Follow this with a piece of blank leader tape.

The tape recorder as a drummer

Music practice with someone playing drums is ideal particularly for popular music but a drummer complete with kit may not be available. You can, however, purchase drum rhythm records which contain 32 bar sets of all kinds of popular rhythms, i.e. waltz, foxtrot, tango and Latin American rhythms, etc. The separate tracks all have 'count-in' beats extra to the 32 bars and such records are made and sold by Ad Rhythm Limited (details in Appendix). With a tape recorder one can, however, copy the discs and extend the fixed 32 bar passages into two or three times 32 bars plus intro's and endings. The method is to copy enough of the requisite rhythm track on to tape and then cut and splice the tape to produce continuity of the recording. With practice it is quite easy to cut a tape and splice again exactly 'on the beat'.

The tape recorder as an accompanist

Here is another way in which to use a tape recorder but which applies mainly to keyboard instruments. Try recording only the accompaniment, left hand in the case of the piano and piano accordion and left hand and pedal bass for organs. This can be replayed whilst you accompany with the right hand thus allowing more concentration on the melody. Don't forget that a count in signal is required before the first beat of the music. This method also offers scope for those who have advanced sufficiently to do a little improvisation on otherwise standard pieces of music, e.g. to practise fill-ins over parts where the melody note holds for a long time or for introductions and endings, etc. First record the accompaniment accurately from first to last beat. Organists may include the bass line. This is now used as the foundation against which the melody and bits of improvisation can be practised. A little practice on these lines can help considerably to improve style and playing generally.

Multi-music recording

Some years ago an American guitarist, Les Paul, started a new vogue in instrumental music by making multiple recordings with a guitar using the one instrument to produce the chord and bass accompaniment, harmony and melody. The basic method was to first record say the chord backing and then re-record this together with the bass line and then repeat the process, adding the harmony and melody, until the orchestration was complete. One of the Les Paul features was to record parts at half the normal recording speed which, when replayed at the proper speed, i.e. at twice the speed recorded, were reproduced one octave higher in pitch and at twice the original playing speed. By this method difficult passages with lots of notes per bar could be played slowly whilst recording but reproduced with speed and accuracy seemingly beyond the capabilities of the most expert performer. The technique is really very simple. Record the accompaniment at say 7½ ips tape speed. Now replay this at half that speed (3¾ ips). The pitch is the same but is one octave lower and the tempo is half the original rate. A complicated melody passage, or one that would have originally been very fast, can now be played slowly but simultaneously recorded with the slowed down accompaniment. Both parts are now replayed at the original speed of the accompaniment, i.e. at 7½ ips. This is of course only one of the 'tricks' of tape recording that can be applied to multi-music. Any musical instrument may be used in the making of multi-music but electrical or electronic instruments are ideal as they can be connected directly to a tape recorder thus eliminating the use of a microphone. The writer frequently uses an electronic organ and electric guitar although rhythm (drum) backing is usually produced by recording via a microphone (see Fig: 5.7, *opposite page*).

The Art of Multi-track

The term 'multi-track', when applied to multiple recordings made with domestic or small semi-professional tape recorders, is not really a strictly accurate one. In the professional field multi-track usually means a number of separately recorded tracks on a wide tape as in Fig: 5.5 and using a recorder like that shown in Fig: 5.1. With domestic tape recorders multi-track means recording from one track to another on the same tape and at the same time adding new material to the previous recording. There are many recorders now available with the track-to-track facility, in fact most stereo models have this feature. The alternative and in fact a much better one, is the use of two tape recorders whereby transfer is made from one to the other with new material being added each time. This is the method employed by Wout Steenhuis whose multi-guitar recordings are so well known (see Fig: 5.8, *opposite page*). He normally uses two semi-professional stereo recorders and his procedure is to copy from one to the other using a technique that will be described later. What is most important, however, is that the recording equipment has a low noise performance because noise can build up, through successive recordings, to a highly undesirable level. Tape speeds play an important part if the double speed technique is to be used and for this the machine must have at least two operating speeds of either 3¾ and 7½ ips or 7½ and 15 ips.

The following techniques can be applied equally to single recorders with track-to-track facility or to the use of two tape recorders when recording from one to the other. The order in which the various parts are recorded depends largely on the musical instrument(s) being used and on the music itself and needs careful thought. The process of re-recording introduces a certain amount of distortion so the initial parts should be those of secondary importance such as drum rhythm, chord and bass parts and counter-point, in that order. The melody parts are almost always recorded last. As an example let us take a four part recording using a piano and an acoustic (non electric) guitar. To begin with a microphone must be used and any monitoring must be carried out with headphones. Assume now the following musical arrangement—guitar chord accompaniment, bass part (from low register guitar strings) piano taking melody (first chorus) guitar taking say introduction and melody in second chorus. The basic chord accompaniment from guitar will play through the whole piece and should be recorded first. The bass line is added next and as these parts are 'background' they must appear in the final recording at appropriate level. Try to imagine you are recording all the parts as a group of instrumentalists would play them, and consider how they would be balanced, one against the other, in terms of loudness. Balance of this nature is just as important in multi-track but has to be done with the volume control. So first the guitar chord accompaniment will be recorded at full recording level. The bass part is then recorded at a level appropriate to the chord part. The level of both is then reduced to balance with melody parts when they are all recorded together.

'One man band' recordings of this nature can sound very pleasing and many interesting effects such as reverberation and tape echo can be introduced. However, musical arrangement and timing, etc., is most important. It is difficult, if not impossible, to produce tasteful multi-track music by haphazard arrangements 'played by ear', or made up as you go along. Those who do 'play by ear' would soon discover that precision timing is most essential. Coming in half a beat too soon or too late will throw the whole thing out completely. This leads now to the use of count-in signals without which it is virtually impossible to commence playing on the first beat. Any kind of signal will suffice such as a voice count of one, two, three, four, or a tapping sound of four beats preceding the actual first beat of the score. This count-in signal can be cut from the tape when the whole recording has been completed. For those who require drum rhythm track the Ad Rhythm records mentioned earlier can be used very effectively and can be re-recorded on tape to make up complete rhythm tracks for intrumental backing.

Tape Loop Rhythms

These were mentioned briefly in Chapter 4 and can be employed successfully in multi-track music. The general procedure is to record percussive sounds, cut pieces of these from the tape and splice together in an endless loop. The loop is then copied for the requisite number of bars. Some experiment will reveal that quite fascinating rhythms can be produced this way and there is no reason why additional sounds should not be added during the copying. Loops can also be recorded at one speed and played at another to produce the required tempo. The writer employs a tape desk equipped with different sized capstans for this purpose as shown in Fig: 5.9, *following page 144.*

Multi-Track Equipment

Domestic and semi-professional tape recorders intended for multi-track purposes will normally have built-in switching for using the facility and instructions issued with the recorder will explain how it is used. Normally one should never expect to do more than three or four re-recordings which means that the first recording made will be copied three or four times. Beyond this the quality will suffer and the noise level will become too high. Much the same applies to recording by copying from one tape recorder to another.

Those who contemplate very serious multi-track would either have to employ an expensive professional multi-track recorder or utilize the stereo recorder-to-stereo recorder technique that is used by Wout Steenhuis and many other professionals. This also entails using an external signal mixer and a monitoring amplifier and speakers, etc., in an arrangement (used by the writer) similar to that shown in Fig: 5.10. In this two stereo half-track recorders are used both of which have high level signal inputs that can be fed from an external signal

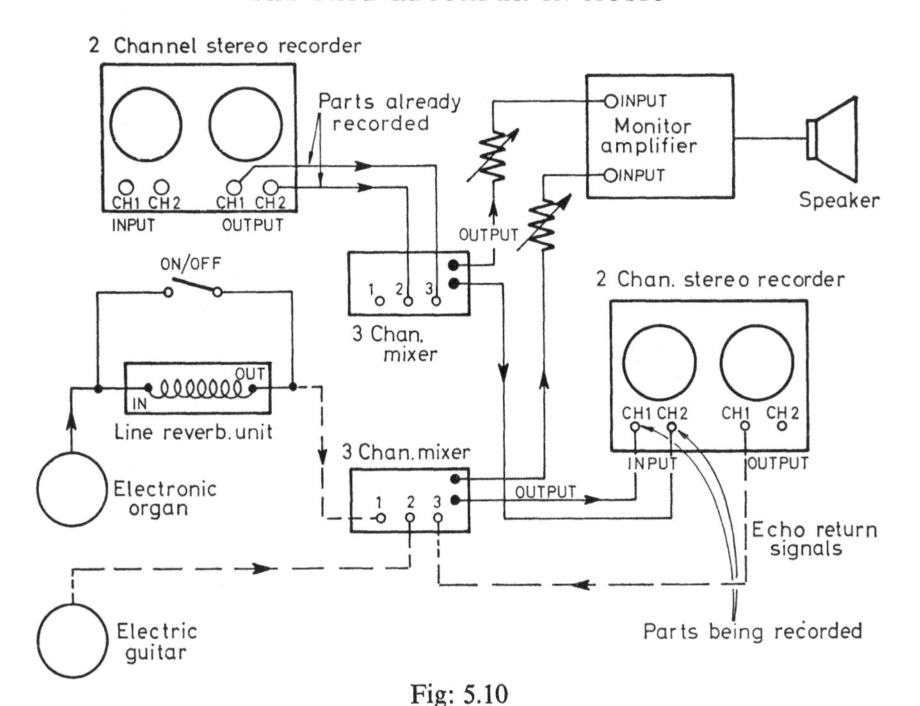

Fig: 5.10

A typical arrangement using two stereo tape recorders for multi-track music with electronic organ and electric guitar.

mixer. The procedure is as follows. The first part is recorded on one track of a stereo recorder. The output from this (for replay) is taken via a signal mixer to one of the channels on a second stereo recorder. The next part is recorded on the remaining channel of the second recorder. This means that parts one and two are on separate tracks on the same tape. These are then replayed together via a mixer so that each can be balanced appropriately and re-recorded on one channel of the first recorder. The next music part is recorded on the remaining track of that machine. The process is repeated accordingly. It is essential that all parts being recorded and re-recorded are audible so in addition to the recording and signal mixing equipment monitoring amplifier and loudspeakers (or headphones) are necessary.

A signal mixer suitable for use with domestic tape recorders is not a very expensive item but is essential when working with one or more

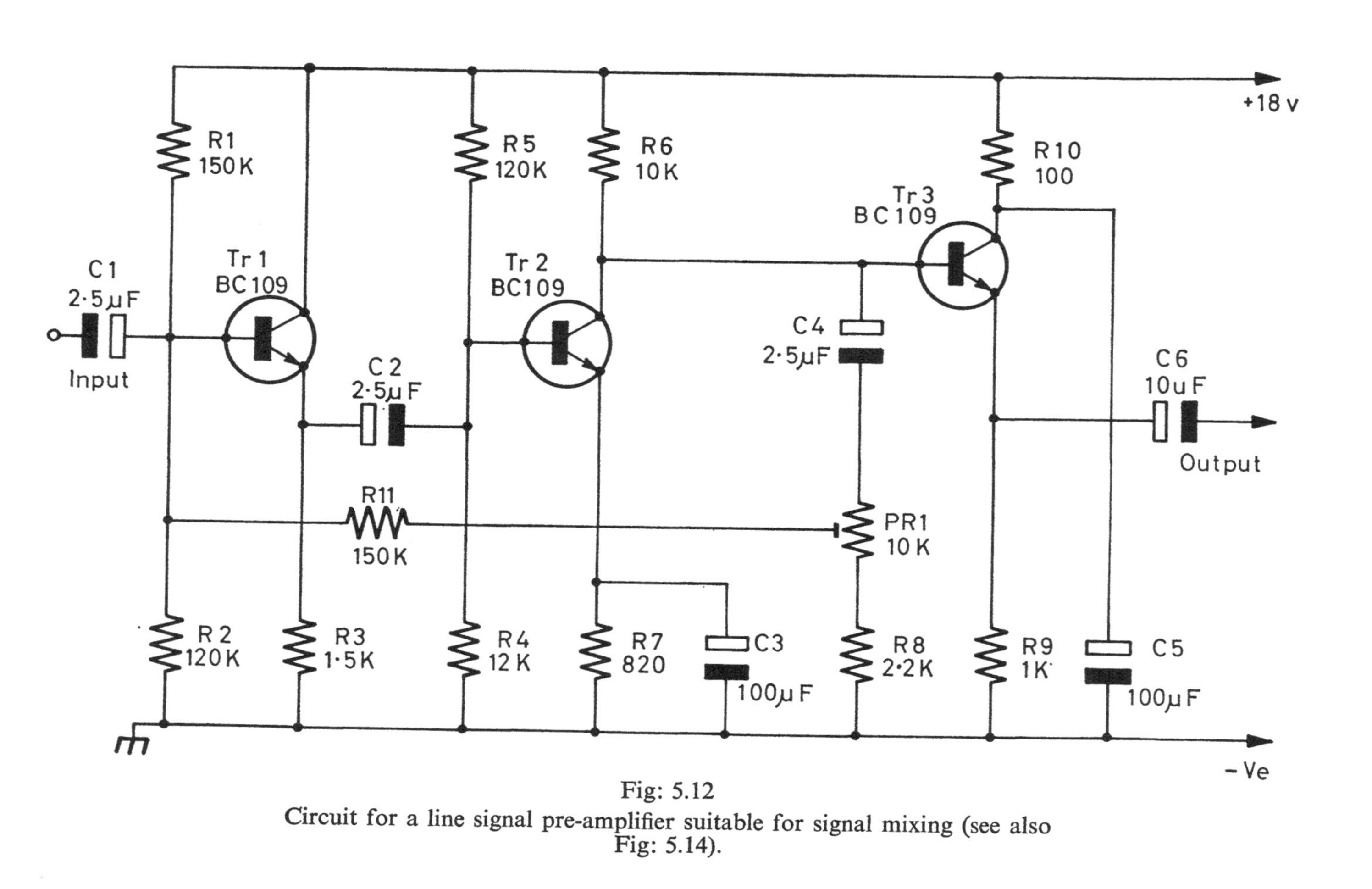

Fig: 5.12
Circuit for a line signal pre-amplifier suitable for signal mixing (see also Fig: 5.14).

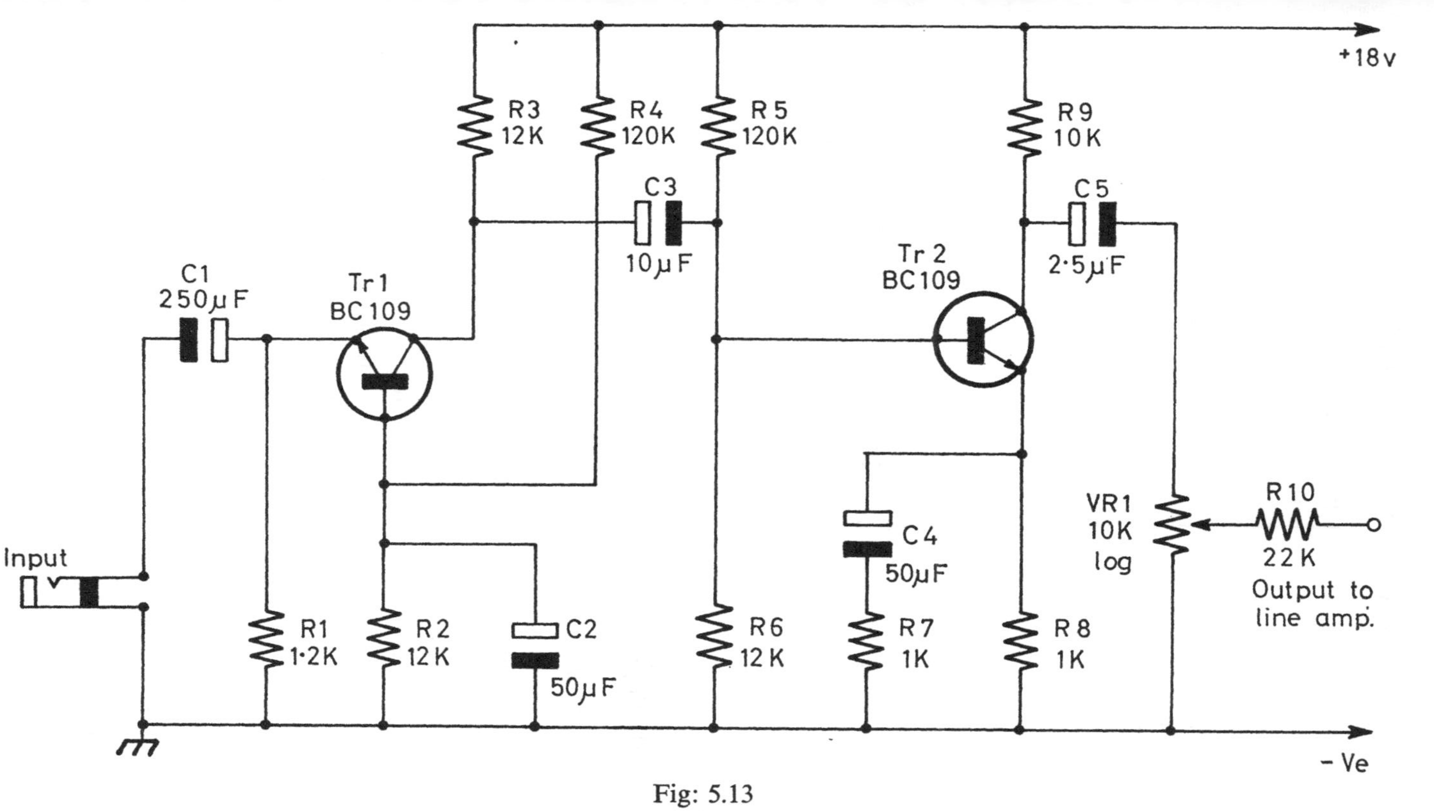

Fig: 5.13

Pre-amplifier for microphones of between 25 and 200 ohms impedance and for use with the line amplifier in Fig: 5.12.

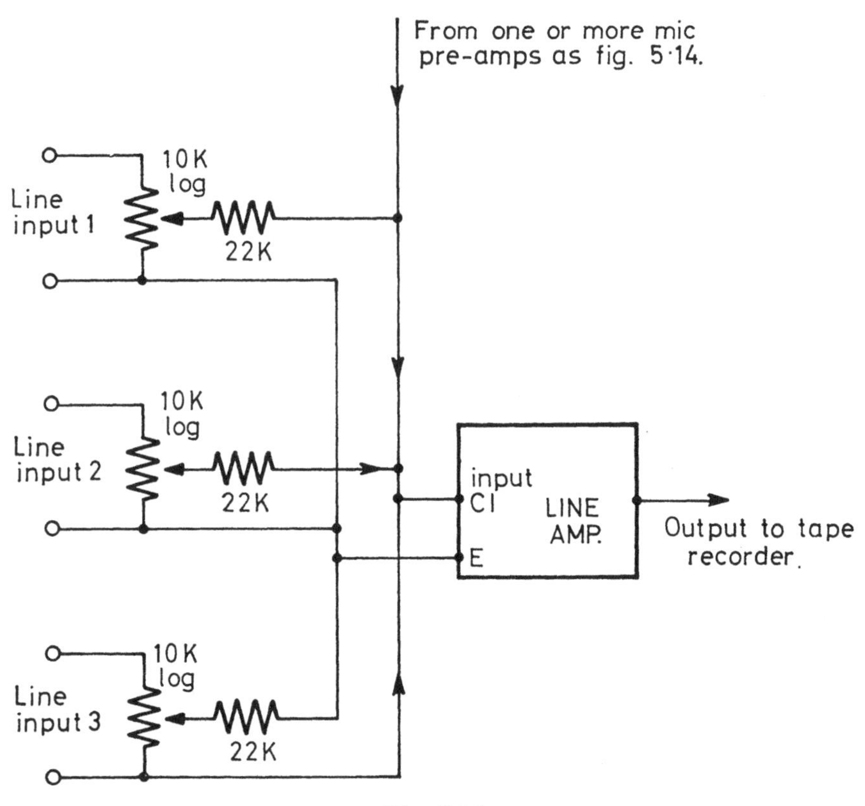

Fig: 5.14
Passive mixing network for use with line amplifier shown in Fig: 5.12.

tape recorders and particularly those without built-in facilities for signal mixing. If multiple recordings are to be made with acoustical musical instruments then a microphone has to be used in which case the signal mixing unit must have one or more inputs for microphones and one or more high level inputs, i.e. inputs designed to accept signal levels likely to be obtained from the replay pre-amplifiers of a tape recorder and which may be as high as 1V rms.

A mixer of this nature, made by Grampian Limited is shown in Fig: 5.11, *following page 144*. The circuits given in Figs: 5.12 and 5.13 can be used as the basis for a fairly versatile mixing unit with one or more microphone and line inputs. The line amplifier (Fig: 5.12) can be used with a number of passive mixing networks for line input as shown in Fig: 5.14 to which may be added the outputs from one or more microphone pre-amplifiers as in Fig: 5.13. The microphone pre-amplifier has its own gain control and series resistor (R10 22K) so its output, from R10, can be connected directly to the input at C1 of the line amplifier. Each of the inputs to the line amplifier, as in Fig: 5.14 are suitable for up to 1V or more of signal as would be obtained from the line output of a tape recorder. The line amplifier itself has an output of 1V rms for 100mV input at C1 but its overall gain can be adjusted by the pre-set resistor (PR1 10K). The microphone pre-amplifier is suitable only for the direct connection of low impedance (25 to 200 ohms) microphones.

Reverberation and Echo Units

These are dealt with fairly extensively in Chapter 3. Various designs for home construction have been published in the technical press but most have a performance somewhat inferior to commercially made units. Good spring line reverberation units are difficult to obtain and mechanisms suitable for continuously running tape loop or tape drum echo systems are just not available to the amateur constructor.

CHAPTER SIX

Electronics in Music Reproduction

The term 'hi-fi', which means high fidelity, is now universal and indicates that the reproduction of music and sound is to a standard of quality resembling as closely as possible that of the original. High fidelity begins in the recording or broadcast station studios where every effort is normally made to ensure that the quality of the original is not distorted by the addition of unwanted harmonics (harmonic distortion) or by the inclusion of unwanted noise, i.e. noise not relevant to the original. It is equally important that the dynamic (relative loudness) range and frequency response are both maintained as near to that of the original as possible. The media for conveying the original to the listener will be either radio, gramophone, disc or magnetic tape and if it is assumed that these are above reproach then 'high fidelity' in the home depends entirely on the equipment used for reproduction from these sources. The sources themselves may not always be perfect but the listener has no direct control over them. One can only ensure that little or no further deterioration of signals from these sources will occur simply by using the best possible record transcription unit, tape recorder or radio tuner and amplifier and loudspeaker(s) available. Equipment capable of fully meeting this demand is expensive and would in fact amount to the finest that could be purchased. However, a certain amount of signal deterioration can be accepted without spoiling reproduction to the point where it becomes unacceptable as hi-fi. This means that equipment 'cheaper than the finest' can be used with the knowledge that little will be lost and what you will not hear will not be missed anyway.

The Hi-Fi System

A really complete system for hi-fi stereophonic reproduction consists of a twin channel amplifier or an integrated radio tuner amplifier, a pair of loudspeakers, a radio tuner (if not integrated with the amplifier) a record transcription unit with stereo pick-up and a twin channel (stereo) tape recorder or tape record/replay unit. Normally a tape record/replay unit is used as this operates directly with an external amplifier. A choice of the best on the market could run to well over £500 for a complete system but on the other hand quite acceptable reproduction can be obtained from a system costing less than half that amount. For anything costing even less the quality of reproduction might still be very good but perhaps only on the border of real hi-fi.

Performance Specifications

Equipment labelled hi-fi should perform to given specifications, which in Germany for example, are derived from the DIN (Deutscher Industrie Normenausschus) 45,500 standards for all hi-fi reproducing equipment. In Great Britain we have no special standard like the DIN 45,500 but instead adopt various standards for audio equipment performance laid down by the British Standards Institute and other organizations. It is on these that most of the performance specifications of British made equipment are based. Our standards are in fact a little higher than those set by the German DIN 45,500 which is also adopted by most other European countries and Japan.

Each system that goes to make up a high fidelity reproducing system has its own particular performance specification and for an amplifier for instance this would include a specified power output, frequency response, distortion factor and minimum hum and noise level, etc., as well as specified limits of control over frequency response and specified input signal levels, etc. With radio tuners the performance specification is concerned with wave band coverage, sensitivity, noise level, image signal rejection, etc. Tape recorders necessitate both electrical and mechanical performance specifications which usually stipulate frequency response, hum and noise level, etc., as with amplifiers and on the mechanical side may set down the nominal tape speeds, maximum permissible long and short term deviations from the nominal tape speeds, tape rewinding time and maximum and minimum spool sizes, etc. The specifications for record transcription units relate running speeds and deviations and factors concerned with the pick-up arm balance and tracking, etc. Pick-up cartridge performance is concerned mainly with frequency response, signal output, crosstalk (if stereo) tracking weight and compliance, etc.

Development

High fidelity sound reproduction did not really begin to develop until after 1945 although many efforts were made prior to this to improve the quality of reproduction from radio and disc records. The introduction of negative feedback in audio amplifiers as a means of obtaining more linear and wider frequency responses for example came into use around 1934 and has since remained one of the most important design features of all modern amplifier systems. Aside from improving frequency response, negative feedback assists considerably in the reduction of distortion and phase shift and can be used to control the gain of an amplifier.

Stereophonic reproduction is now accepted as part and parcel of hi-fi and although work in this field was initiated by A. D. Blumlein of EMI, in 1929, the major developments in systems and equipment for stereophonic sound reproduction did not really begin until after 1945. It is interesting to note that although A. D. Blumlein produced 78 rpm stereo disc records in 1933, stereo LP records as we know them today did not appear until some time after the first stereo tape records which became available in about 1955. More recently stereophonic broadcast programmes have been introduced. Stereo reproduction at home can now be obtained from magnetic tape, disc and radio.

The latest development, which is an extension of stereo, is one that utilizes more than two channels. Stereo normally makes use of two separate signal channels, from microphones to loudspeakers, in order to convey the spatial effect, i.e. the spread of sound that would normally be heard in the studio or concert hall. Systems using more than two channels are now being employed to present to the listener not only the width of the sound image but also the depth which includes the residue of sound that would be heard by a listener in a studio or concert hall from the immediate sides and from the rear.

Most of this sound is due to reflections from walls, etc., but is considered as part of the whole sound image.

Amplifiers

The nucleus of a high fidelity reproducing system is the amplifier and its primary function is to receive and amplify the signals from various sources such as magnetic tape, disc or radio, to sufficient power to drive the loudspeaker(s). A hi-fi amplifier will normally have controls for varying its overall frequency response, for adjustment to volume, stereo balance (in the case of stereo amplifiers) and for selection of the signal sources connected to it. Practically all hi-fi amplifiers intended for domestic use are stereophonic which means that the one unit will in fact contain two identical amplifiers, one for each channel. A typical hi-fi amplifier is shown in Fig: 6.1, *following page 144*. A more recent trend has been to incorporate an AM/FM tuner with the amplifier and such units are known as integrated tuner/amplifiers. The general performance requirements of the amplifier section, however, remains the same.

Loudspeakers

The main requirement of a high fidelity loudspeaker is that it must be capable of handling the full power from its driving amplifier and do so without introducing an undue amount of distortion. The frequency response must be as uniform as possible over the entire audio frequency spectrum. It is difficult to produce a single loudspeaker that will cater for all these requirements by itself and common practice is to employ two or more speaker units each designed for optimum performance over a given frequency range. A complete hi-fi speaker unit may therefore consist of a bass unit covering the low frequencies from about 20Hz to 5000Hz and a high frequency or 'tweeter' unit covering from about 500Hz upwards. An extension of the idea is to use three speakers similar to the arrangement shown in Fig: 6.2, *following page 144*, one for the bass, one for the midrange frequencies (about 400 to 5000Hz) and one for the high frequencies. Each speaker is isolated frequency wise from the others by filters known as crossover networks. Proper bass reproduction, however, requires that the bass speaker unit be mounted in an enclosure or cabinet designed to be resonant over the greater portion of the low frequency range. Without the enclosure the low frequency speaker would be very inefficient and the result a poor bass response. As it is desirable to have all the speaker units out of sight, i.e. inside a cabinet, this in itself may be used, providing it has the appropriate dimensions, as a resonator for the bass speaker unit.

Radio Tuners

As already mentioned these are sometimes integrated with amplifiers in which case the unit becomes known as a tuner/amplifier. One trend of late has been to build a tuner/amplifier within the same cabinet containing a record transcription unit and thus, with the exception of the speakers, provide the essentials of a hi-fi system. Radio tuners are fairly standardized affairs generally covering the VHF FM local broadcast band and the medium wave broadcast band. Most now incorporate a stereo multiplex decoding unit for separating the signals for each stereo channel from multiplex FM stereo broadcasts. The multiplex system is one which is compatible, i.e. those with receivers not fitted with a stereo decoding unit will receive the stereo transmission monophonically. Recent developments in VHF FM tuner circuitry have resulted in signal to noise and distortion factors almost equal to those obtained with a top quality amplifier.

Tape recorders and record transcription units

Neither could be called new developments but over the years improved design has resulted in performance that is acceptable for high fidelity. The record transcription unit is perhaps one of the most important hi-fi signal sources and the model shown in Fig: 6.3, *following page 144*, is typical of its kind. It consists essentially of a heavy record turntable driven by a constant speed motor and both are engineered to a precision that ensures a minimum amount of periodic speed fluctuation (wow) and minimum amount of mechanical vibration. Low frequency mechanical vibration can become amplified via the record pick-up and is an effect known as rumble.

The pick-up arm is also a carefully engineered device and its essentials are light weight, accurate tracking and precise balance.

Accurate tracking stems mainly from the shape of the arm which ultimately determines the angle at which the pick-up will track with respect to the record grooves. Balance is usually obtained by counter weighting which is adjusted when the pick-up cartridge is fitted to obtain a given pressure by the cartridge stylus onto the record. Most pick-up cartridges are now stereo magnetic types similar to that shown in Fig: 6.4, *following page 144*. As far as hi-fi is concerned these have a sufficiently wide frequency response, coupled with fairly low cross-talk and with an accurately aligned and balanced p.u. arm are capable of providing an output signal with relatively little distortion.

The conventional reel to reel tape recorder is also a fairly standardized piece of equipment and there are many that meet the requirements for hi-fi with regard to signal to noise ratio, wide frequency response, low distortion and adequate mechanical performance. Most tape recorders available are for stereo operation but many, known as tape record/replay units, do not have built-in loudspeakers, and are intended for use with an external amplifier and loudspeakers. For this reason they are ideal as the tape recording and replay source for hi-fi reproduction and will operate with practically all hi-fi amplifiers. Track systems and tape speeds are dealt with in Chapter 5. It has been advocated that the cassette and/or cartridge tape system will one day supercede the reel to reel recorder and as a pre-recorded signal source even take over from the gramophone disc. This is not likely to be for some years, although pre-recorded music cassette and cartridges are becoming popular, particularly now that noise reduction can be applied and which brings the noise from pre-recorded music tapes down to a level comparable with that from modern LP gramophone records.

Dynamic Range

One of the usual concerns of hi-fi enthusiasts is how much power the amplifier will deliver and many are under the wrong impression that hi-fi must be loud and therefore, the power output of the amplifier must be very high. High power output does of course permit very loud reproduction but its real usefulness is connected mainly with another factor called 'dynamic range'.

In most music there are parts where the sound level reaches a certain peak loudness and parts where it may fall to a certain minimum loudness, i.e. very quiet passages. The range of loudness between very quiet and very loud is known as the 'dynamic range'. The reason for using an amplifier having a fairly large power output is to reproduce this dynamic range as closely as possible but on a scale more suitable to the confines of the average living room. For example, the sound power produced by a large orchestra in a concert hall may be equivalent to around 100 watts and this would be deafening in a living-room. If we bring this down to living-room level then a power of around 15 to 20 watts, or even less, is sufficient to create the same impression of loudness.

The loudness or dynamic range of all sound is, however, much wider than that of musical instruments alone. The dynamic range of all sound begins at a point known as the 'threshold of hearing' which is where sound first becomes perceptible to human ears. The very loudest sounds occur at a point where the intensity is so great that it becomes painful to listen to. This level is known as the 'threshold of feeling'. Loudness is usually expressed in 'phons' and the loudness level of a sound, in phons, is numerically equal to the intensity level in decibels of a 1000Hz pure tone which is judged by a listener ,with normal hearing, to be equally loud. The reference intensity is 10^{-16} watt per square centimetre which is near to the value of the threshold of audibility for a 1000 Hz pure tone. Phons can only be expressed directly in decibels when the 1000Hz reference is used because loudness is a function not only of intensity but also of frequency as shown in Fig: 6.5. However, at frequencies around 1000Hz the dynamic range of sound can be taken as about 120dB. When we are dealing with reproduced music, however, the lowest sound level does not begin at the threshold of hearing but at a point much higher and in the studio this is the ambient noise of the studio itself. The total dynamic range available at the microphone may therefore be not more than about 100dB. Between the microphone and the listener at home there is the reproduction chain which includes not only the media for conveying the signals to the listener but also a number of amplifiers. All of these can introduce noise and although this may emerge from the end of the reproducing chain at a constant level, it does nevertheless set a point which determines the lowest level of

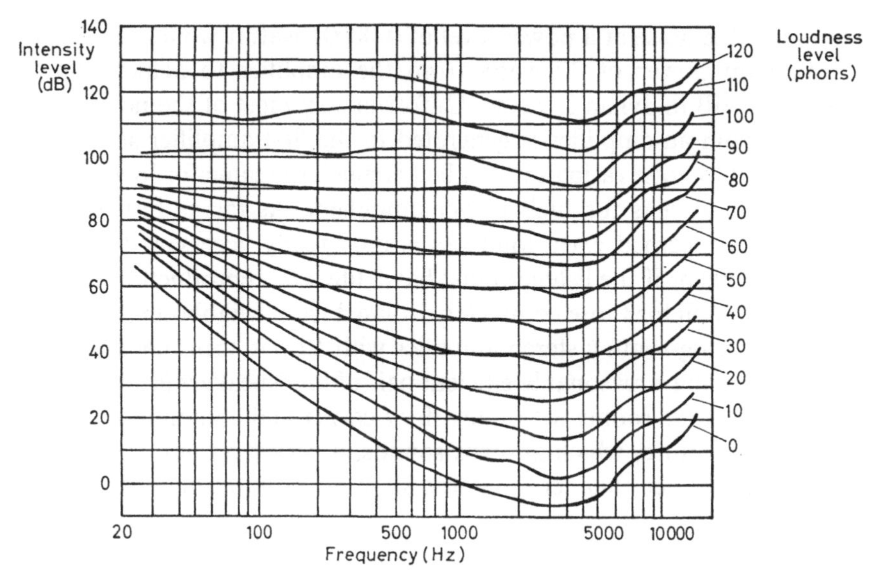

Fig: 6.5
Contours of equal loudness (*Fletcher and Munson*).

wanted sound that can be heard from the loudspeakers. That point is where the level of wanted sound meets that of the noise. The noise level therefore, sets what might be called 'the threshold of sound reproduction' and which is in fact the starting point of the dynamic range of sound from a reproducing chain. Any signals concerned with reproduction, i.e. signals stemming from the original, that are below the noise level of the system, are just not heard. This narrows the overall dynamic range of a reproducing chain quite considerably and the best that can be achieved without undue distortion is not greater than about 70dB. The average is more likely to be between 50 and 60dB, depending on the signal source. For example if we accept the power of the noise level at the output of the final link in a reproducing chain, i.e. at the output from the amplifier driving the loudspeakers, as say 0·00001 watt, then 10 watts of power from the amplifier would represent a dynamic range of 60dB. It might seem reasonable to suppose that increasing the power output, without increasing the noise level, would result in a wider dynamic range. This is in fact true but the increase is not very great. For example, if we retained the noise at a power level of 0·00001 watt it would require a maximum power of 20 watts from the amplifier to achieve only another 3dB in dynamic range. Unfortunately most of the noise that comes from an amplifier is generated in the first amplifier stages so the signal to noise ratio as it is called is much the same for low power amplifiers as it is for high power amplifiers. As far as hi-fi in the home is concerned extra loudness is only really necessary for very large rooms or if the listener must have 'life size' reproduction which is all very well for those completely isolated from neighbours. For the average living-room a stereo amplifier with 10 watts rms per channel output driving a reasonably efficient pair of loudspeakers is adequate. For fairly large rooms with thick floor carpet and containing lots of padded furniture, which absorbs sounds, a power output of around 20 watts rms per channel for stereo should be quite sufficient.

Distortion

Low distortion from amplifiers is equally important as a low unwanted noise level. There are two major forms of distortion one of which is intermodulation, an effect due to one frequency becoming modulated by another causing the production of spurious tones. In modern hi-fi amplifiers this form of distortion is usually very low, so low in fact, that it is rarely cause for concern. The other and most common form of distortion is that arising from the generation of harmonics within the amplifier and which is called harmonic distortion. Again the amount likely to be obtained from a well designed high quality amplifier is usually very low, much lower in fact than would be obtained from most available signal sources. Harmonic distortion is usually quoted as a percentage related to the rms voltage developed across the output load of an amplifier when its power output is at maximum rms value in watts. For example one might find that the performance specification of an amplifier quotes total harmonic distortion (THD) as 0·1% at 1000Hz related to 10 watts rms power output. The 0·1% is the total rms voltage of all the harmonics that remain after removing the fundamental (1000Hz) test signal. If the amplifier output load is 8 ohms and the rms power is 10 watts, the rms voltage across the load will be $\sqrt{Po \times RL}$ where Po is the power output in watts and RL the output load resistance.

Fig: 5.7 Equipment used by the author for multi-track music recording. Electronic organ, electric guitar, drums, two tape recorders, signal mixer and monitor amplifier.

Fig: 5.8 Wout Steenhuis, popular multi-track guitarist, uses two stereo recorders for multi tracking as outlined in the text and in Fig: 5.10 overleaf.

Fig: 5.9 A Brenell tape deck with various sized capstans used for playing tape loops at different speeds.

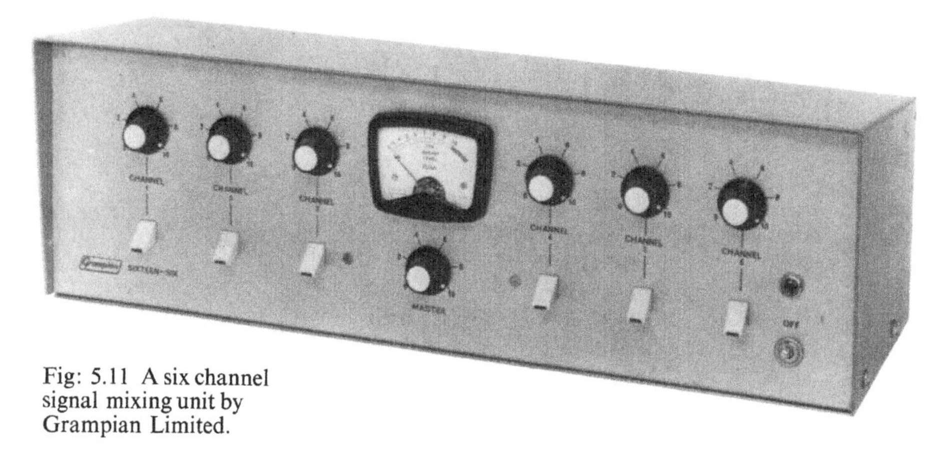

Fig: 5.11 A six channel signal mixing unit by Grampian Limited.

Fig: 6.1 A typical hi-fi amplifier. The SV140 by Grundig (GB) Limited.

Fig: 6.2 The KEF Cadenza loudspeaker.

Fig: 6.3 Hi-fi record transcription unit by Thorens.

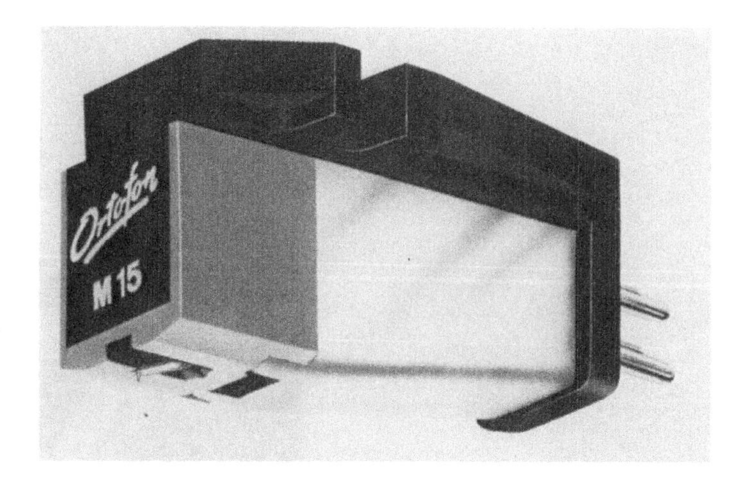

Fig: 6.4 Ortofon magnetic pick-up cartridge.

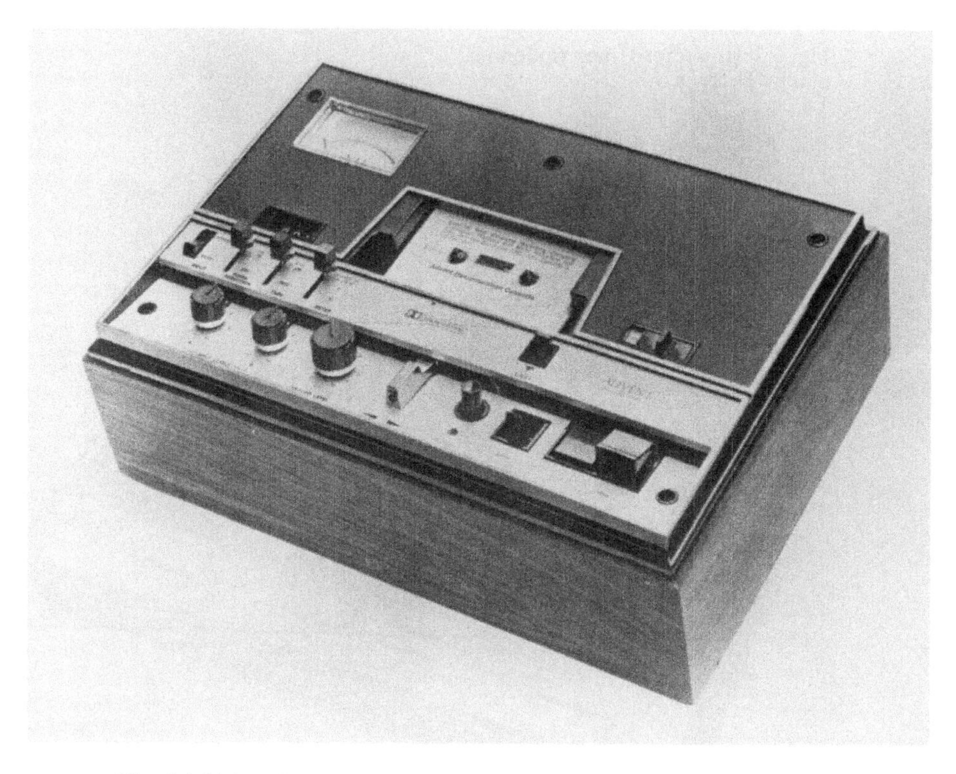

Fig: 6.6 'Advent' cassette tape player with built-in Dolby B noise reduction system.

The rms voltage will be $\sqrt{10} \times 8$ or approximately 8·9V or 8900mV. Now, with the aid of a distortion factor meter the fundamental test signal is filtered out and what is left is the total residual rms voltage due to all the harmonics. If this amounted to 8·9 milli-volts then the percentage of harmonic distortion would be $\frac{8{\cdot}9\text{mV}}{8900\text{mV}} \times 100\% = 0{\cdot}1\%$

Intermodulation distortion is also expressed as a percentage and related to power output in much the same way as for harmonic distortion. In this case, however, two test signals each at different and widely separated frequencies are used. The intermodulation distortion percentage is arrived at by filtering out the two test signals and computing the rms signal left as a percentage of the full rms output from the amplifier. As with harmonic distortion the percentage is of the total remaining signals but these can be evaluated separately with the aid of a wave analyser.

Frequency Response

A poor frequency response is a form of distortion but unlike harmonic and intermodulation distortion is less likely to irritate the ear. A wide uniform frequency response is, however, essential to high fidelity reproduction in order that higher frequency harmonics associated with music are faithfully reproduced (see Fig: 1.1 Chapter 1). Achieving wide frequency response is a problem that has been virtually eliminated and it is not uncommon to find amplifiers with a response extending from about 10Hz to almost 100,000Hz, far beyond the range of natural hearing. However, uniform response is not quite sufficient and it is equally important that an amplifier is also able to maintain uniform power output over at least the audio spectrum. This too is now fairly easy to achieve and power bandwidths to within 1dB from about 20Hz to 30,000Hz are commonplace.

Hum and Noise

In the early days of high fidelity sound reproduction one of the biggest problems that faced designers of audio equipment was to keep hum and noise down to an acceptable minimum level. The ideal, absolutely no hum and noise at all, has not yet been achieved. Noise refers to what is commonly called random noise, sometimes known as white noise and is a continuous signal that presents itself audibly as a hissing sound. Unfortunately noise of this kind has elements of random frequency, phase and amplitude and because of this is impossible to eliminate completely. The best that can be done is to ensure that transistors and other components used in high quality audio equipment are such that each contributes an absolute minimum of noise. The noise level from present day amplifiers and even tape recorders does not amount to very much and a noise level of −60dB with reference to wanted signal voltage is the rule rather than the exception. This means that the noise level is 1000 times less than the signal voltage. However, amplifiers or other equipment at home are not the only contributors of noise. Beginning at the broadcast or recording studio, noise can be added to the signals from the microphone by the microphone amplifiers, studio mixing equipment and tape recorders, etc., and in the case of broadcasting by the transmission equipment as well. Any noise produced by equipment at home, i.e. the reproducing equipment, will be added also.

However, random noise is not the only unwanted signal, aside from spurious harmonics and signals due to intermodulation. All the equipment in the complete chain from studio microphone through to the final amplifier stage in home equipment is operated from 50Hz mains supplies. It would just not be practical or economical to use batteries. When mains supplies are used as a power source for audio equipment they can introduce hum onto wanted signals in many different ways. As random noise and hum are both unwanted signals it is usual to evaluate them together, particularly where domestic hi-fi equipment is concerned. Performance specifications therefore usually quote 'hum and noise', minus so many dB's, relative to rated power output. Hum is almost impossible to get rid of once it becomes part of a wanted signal and any attempt to filter it out can seriously interfere with the low frequency response of the equipment. Obviously everything possible is done by manufacturers to prevent hum by the use of screening, etc., and by ensuring that 50Hz ripple (unsmoothed rectified voltage from the mains supply) is kept to an absolute minimum. It is the hum and noise level, however, that sets the lowest level of the dynamic range in any sound reproducing system and unless some new technology should make it possible, completely hum and noise free audio equipment just cannot be produced. The alternative is some means of suppression and/or reduction by electronic means and a recent breakthrough in this direction has been made by Dr Ray Dolby an American audio engineer whose noise reduction system made considerable impact in the field of sound recording, broadcasting and reproduction. The Dolby system is such that, although it can only be applied in certain ways, it has proved extremely effective. It is useable where it is most needed, e.g. in any transmission system where audio signals are passed through a carrying medium such as tape recordings, radio broadcasting and land lines, etc., and which because of their nature can introduce unwanted noise. The Dolby noise reduction system is particularly advantageous in tape recording although for this application there are two systems. The first, known as the A system is essentially for professional studio use and the second, known as the B system is for home use. The A system is capable of reducing unwanted noises such as hum, dc modulation noise and tape print-through and operates over the whole audio frequency spectrum. The B system operates in much the same way as the A system by increasing the level of low level signals during recording and reducing the level during replay. However, the B system operates in one frequency band instead of the four separate bands covering the audio spectrum used in the A system. The single band used in the B system extends above 2000Hz and achieves a maximum noise reduction of 10dB from 5000Hz upward. This does not take care of hum, etc., like the professional A system but does deal with the most objectionable unwanted signal, i.e. noise due to the tape recording and replay process and which is usually known as tape hiss. The B system is relatively simple and therefore, less expensive and is small enough to be built into a tape record replay unit like the tape cassette model shown in Fig: 6.6, *following page 144.* Dolby B system units are also available for use with standard reel to reel tape recorders but one of the primary applications for this system is in tape cassette and/or cartridge record and replay units. Pre-recorded music tape cassette and cartridges can be pre-Dolby treated when they are recorded by the manufacturer and then replayed at home in a unit with the B system (as in Fig: 6.6, *following page 144*) to obtain the appropriate noise reduction. The system is really a two part one, the first part being carried out during recording and the second part during playback. However, it relies on the fact that music consists of sounds of different pitch and loudness and may involve pauses where there is no sound at all. The relationship between loud and quiet sounds can be shown, as in the diagram Fig: 6.7A in which very loud music is represented by the vertical lines on the left and very quiet music by those on the right. The level of noise from a tape recording process is represented by the dotted area in Fig: 6.7B. Now if the music levels are superimposed on the noise level as in Fig: 6.7C it can be seen that very quiet music may be as low as, or even lower

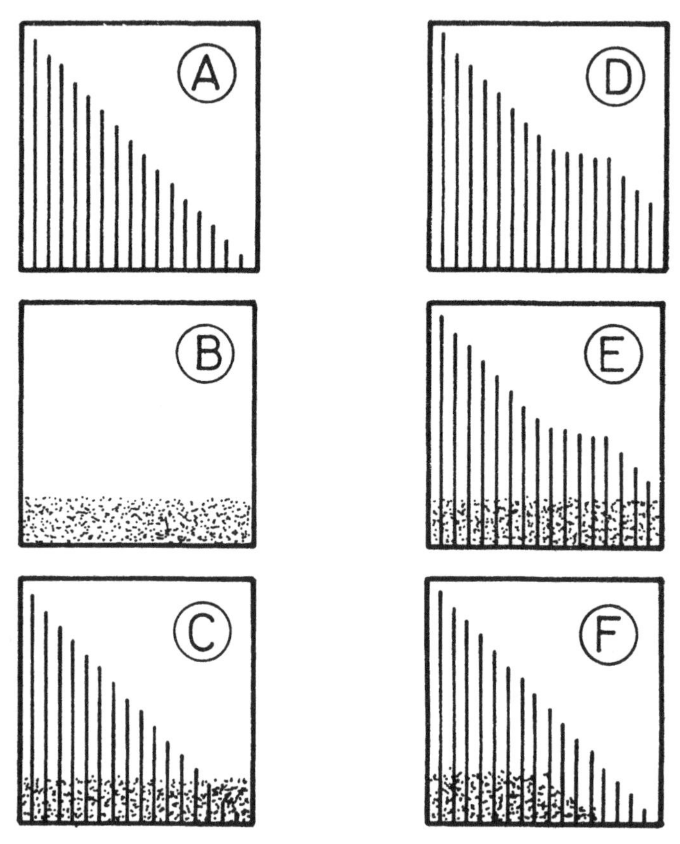

Fig: 6.7
Principle of the Dolby B noise reduction system (see text).

than the noise level. In direct recording and replay any music at a level equal to, or lower than the noise level would be spoiled or barely heard at all. In silent passages only the noise would be audible. What the Dolby noise reduction system does is to first bring up the level of quiet passages as shown in Fig: 6.7D because the system operates on the principle that it can distinguish frequency as well as level. Now, any noise due to the recording process remains constant and as shown in Fig: 6.7E but is now below the level of the quiet passages. During replay the level of the quiet passages are automatically restored to the original and the noise that was also recorded during these parts is reduced as shown in Fig: 6.7F. Although the noise level remains more or less the same with respect to the loud passages it is just not noticed, i.e. as far as the ear is concerned the noise is masked completely by the loud music passages.

Multi-channel stereo

The invention of stereophonic sound reproduction added a spatial effect to reproduced music and consequently greater realism. Sound reproduced by loudspeakers can be spread out to simulate the 'sound width' that would have been heard in the concert hall or studio. However, two channel stereo does not simulate depth, which in music heard live, stems from all the sound that is reflected by walls, etc., and which occurs in any enclosed space like a concert hall. A more recent development, which it is claimed will produce the desired effect in the home, is one most commonly known as 'four channel stereo'. Basically the system is such that the two extra channels are used to carry the sounds that would be heard to the sides and rear of a listener seated in a concert hall or studio. Such sounds consists mainly of reflections from wall surfaces but which may not necessarily be in phase with those heard direct, i.e. from a point directly in front of the listener. The four channel system is used in such a way as to carry not only the direction of reflected sound but also any phase shift that has taken place and which also affects the loudness of reflected and direct sound to a certain degree. By itself the system does not present too many technical problems except that four separate signals channels must be used right through to the listener at home. With tape recording this is fairly easy but if gramophone records are to be used then the record groove must carry four separate signal recordings. In the home the reproducing system will consist of a four channel tape recorder or disc player, a four channel amplifier and four loudspeakers. The loudspeakers are normally placed, two to the front of listeners for direct signals and also for the spatial or width effect as in conventional stereo and two to the rear for the indirect or reflective sound. However, the designers of this system thought it advisable to make it compatible with existing two channel stereo reproducing systems, i.e. so that existing two channel stereo tapes and discs could be played via a four channel system and the signals modified so as to simulate the indirect as well as the direct sounds, i.e. to produce sound depth as well as width. Experiments have also been made with four channel stereo radio.

Whether the system will become as standardized like two channel stereo, or even replace it, remains to be seen. It has been viewed with some scepticism and the most questionable point is concerned with just how authentically can such a system re-create the acoustics of a concert hall or similar environment within the four walls of a living-room. There are still too many imperfections in sound and music reproduction as it stands and it would seem prudent to achieve perfection in quality before embarking on costly systems for improving acoustic environment even supposing the claims made for multi-channel stereo can be substantiated.

ABBREVIATIONS

C	Capacitance
dB	Decibels (one-tenth Bel)
D	Diode
E	Voltage (also expressed as E)
F	Frequency (sometimes Fo)
Hz	Hertz (formerly cps—cycles per second)
I	Current (measured in amperes)
K	Kilo (thousand)
L	Inductance (measured in Henrys Abb: H)
M	Mega (million)
ND	Noise diode
PR	Pre-set potentiometer
pF	Pico-farad
μF	Micro-farad
R	Resistance (measured in ohms)
THD	Total harmonic distortion
TR	Transistor
V	Volts (mV = milli-volts)
VCA	Voltage controlled amplifier
VCF	Voltage controlled filter
VCO	Voltage controlled oscillator
VR	Variable resistance (potentiometer)
W	Watts (measurement of power)

APPENDIX

Bibliography

Audio and Acoustics, G. A. Briggs, Wharfedale Wireless Limited.

Musical Instruments and Audio, G. A. Briggs, Wharfedale Wireless Limited.

Electronic Music and Musique Concrête, F. C. Judd (now out of print but probably available from libraries).

Acoustics, Alexander Wood, Blackie & Son Limited.

The Electronic Musical Instrument Manual, A. Douglas, Pitman Limited.

Transistor Electronic Organs for the Constructor, Douglas & Astley, Pitman Limited.

Hi-Fi in the Home, J. Crabbe, Blandford Limited.

Hi-Fi and Tape Recorder Handbook, G. J. King, Newnes Limited.

Hi-Fi for the Enthusiast, D. L. Gayford, Pitman Limited.

Publications

The following are monthly publications likely to contain articles related to electronics in music, constructional features concerned with electronic musical instruments and hi-fi and tape recording, etc.

Audio, IPC Magazines Limited

Practical Electronics, IPC Magazines Limited

Practical Wireless, IPC Magazines Limited

Everyday Electronics, IPC Magazines Limited

Studio Sound, Link House Publications Limited

Hi-Fi News, Link House Publications Limited

Hi-Fi Sound, Haymarket Publishing Limited

Popular Hi-Fi, Haymarket Publishing Limited

Tape Recording, Anglia Echo Limited

Music Industry (Trade), Music Industry Publications Limited

Keyboard (User), Music Industry Publications Limited.

APPENDIX

Music Synthesizer Manufacturers

R. A. Moog Inc.: UK Agents—Feldon Recording Limited, 126 Great Portland Street, London, W1.

EMS: Electronic Music Studios (London) Limited, 49 Deodar Road, London, SW15.

ARP (Tonus Inc.): UK Agents—F. W. Bauch Limited, 49 Theobald Street, Boreham Wood, Herts.

Drum Accompaniment Records

Ad Rhythm Series: details on request, Ad Rhythm Records, The Broadwalk, North Harrow, Middlesex.

British Amateur Tape Recording Contest

Class for electronic music and multi-track, etc. Open to anyone. Details on request. The Secretary, BATR Contest, 33 Fairlawnes, Maldon Road, Wallington, Surrey.

EVERY PICNIC NEEDS A
SYNTHI
Electronic Music Studios (London) Limited
49 Deodar Road London SW15 Telephone 01-874 2363
New York: 140 East 80th Street N.Y. 10021 U.S.A.
The Synthi Range by EMS

A WORD OR TWO ABOUT EMS...

EMS was founded at the beginning of 1970, to become the first and only company in England engaged in the manufacture of synthesizers. The company was created out of a partnership between EMS's composer/managing director Dr. Peter Zinovieff and design engineer David Cockerell, who had already spent four years together constructing what has now become the most advanced privately owned computerized electronic music studio in the world, located at Putney Bridge.

At that time synthesizers were only available as imports from America and consequently prices were excessive. Clearly there was a need for a British unit. So, together with the assistance of composer Tristram Cary, a small synthesizer was designed for manufacture, and christened 'VCS3' for Voltage Controlled Studio Type 3. It proved not only more versatile but, at £350, considerably less expensive than any comparable equipment. It was an immediate success in music studios, schools, universities and in the serious musical composition field as well as the pop market. It also overcame two major drawbacks of the traditional systems by being highly compact and easily portable, as well as clear and flexible in operation through the use of cordless pin-matrix patching, the system employed in all EMS synthesizers.

The intention of EMS has always been that this commercial venture should finance further development of equipment for the studio, which would, in turn, provide spin-offs for production. An example of this has been the computer technology widely in use at the Putney Studio, which has resulted in the introduction of a unique range of digital sequencers. Up till now sequencers have been characterized by their massive arrays of knobs, requiring tedious setting up, and still giving only short sequences of maybe a dozen events, through the use of simple electronic switching. In comparison the digital sequencer uses a computer-styled memory system, which when directed will remember a sequence of up to 256 control voltages, in the exact pattern that they were loaded, either from a keyboard or any other voltage source.

The Sequencer 256, which forms the heart of the Synthi 100, but is also available separately, is capable of storing six simultaneous but independent tracks of control voltage data, which may each be edited to perfection in a manner that makes it almost analogous with a multi-track tape recorder. The Synthi AKS is a portable synthesizer which includes a 256-event single layer sequencer, making it a powerfully versatile combination at an attractive price. A year after their introduction, these EMS sequencers remain the only available models which employ the time-saving and freedom-giving system of digital memory.

EMS is currently manufacturing the following equipment:

Synthi 100 : Sequencer 256 : AKS : VCS3 : DK1 Keyboard

Modules:

Pitch to Voltage Converter : Random Voltage Generator : Octave Filter Bank : Triple Slew Limiter.

In addition, EMS can also supply complete Computerized Electronic Music Studios.

The range of modules is in the process of being expanded to include all synthesizer devices. Effects pedals and an instrument treatment synthesizer are also on the way.

Introducing the SYNTHI AKS

which **remembers** up to 256 notes as they are played

Ten green apples on the tree; Good for you, good for me
Good for everybody round th' tree

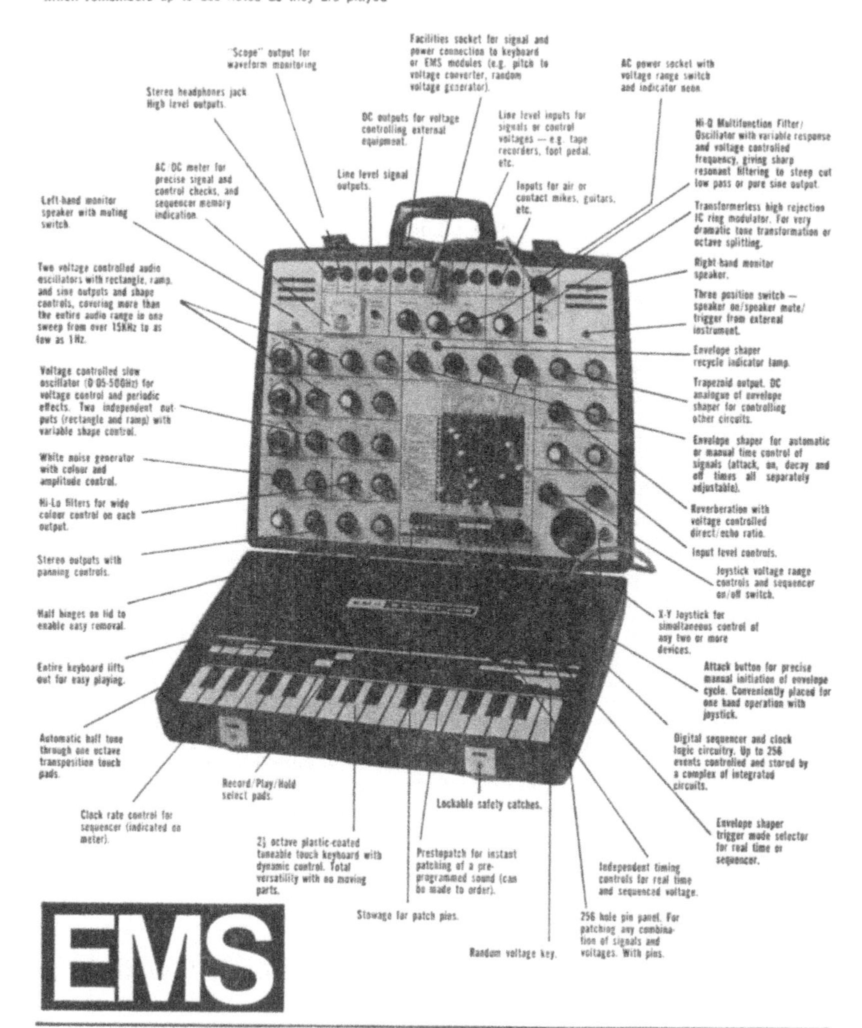

EMS Ltd. 49 Deodar Road London SW15, ENGLAND TEL: (01) 874 2363
EMS Inc. 140 East 80th Street N.Y. 10021, U.S.A. TEL: (212) 628 5200

THE EMS STUDIO AT PUTNEY

Happy with half a guitar?
Now dig this...
We're not knocking 'half guitars'. There are some very good electric 'halves' around. But plug into the Maestro Rhythm 'n Sound Unit and you've got yourself one fantastic whole guitar. Because Maestro is bongos, tambourines, Fuzz tone, brush, percussion, Natural amp, Wah-Wah, Echo Repeat — and a whole lot more in almost infinite permutations.
Flick a tab and you've got a string bass. Tap a footswitch and take it out. Hit two tabs and you're into a new sound no-one's ever heard before. Except you.
How about variable Wah-Wah and Echo Repeat? Or a sensitivity control that gives you only accented notes? The Maestro Rhythm 'n Sound Unit gets it all together. And you make it happen — in a big way.
Then find out about the Maestro Ring Modulator Synthesizer (wow!), too. It's a magic box with a wide range of special instrumental effects. And with a regular microphone you'll make vocal sounds like Planet X. Loud and clear.
RING MODULATOR
Selmer
Henri Selmer & Co. Ltd., Woolpack Lane, Braintree, Essex. Tel: Braintree 2191

Lightning Source UK Ltd.
Milton Keynes UK
UKOW06f1903051015

259919UK00011B/172/P